Lecture Notes in Computer Science 7189

Commenced Publication in 1973
Founding and Former Series Editors:
Gerhard Goos, Juris Hartmanis, and Jan van Leeuwen

Editorial Board

W0192954

Antonio Pescapè   Luca Salgarelli
Xenofontas Dimitropoulos (Eds.)

Traffic Monitoring and Analysis

4th International Workshop, TMA 2012
Vienna, Austria, March 12, 2012
Proceedings

 Springer

Volume Editors

Antonio Pescapè
Università di Napoli "Federico II"
Dipartimento di Informatica e Sistemistica
Via Claudio, 21, 80125 Napoli, Italy
E-mail: pescape@unina.it

Luca Salgarelli
Università degli Studi di Brescia
Department of Electronics for Automation
Via Branze, 38, 25123 Brescia, Italy
E-mail: luca.salgarelli@ing.unibs.it

Xenofontas Dimitropoulos
ETH Zurich
Inst. für Techn. Informatik und Kommunik. Netze
Gloriastr. 35, 8092 Zürich, Switzerland
E-mail: fontas@tik.ee.ethz.ch

ISSN 0302-9743   e-ISSN 1611-3349
ISBN 978-3-642-28533-2   ISBN 978-3-642-28534-9 (eBook)
DOI 10.1007/978-3-642-28534-9
Springer Heidelberg Dordrecht London New York

Library of Congress Control Number: 2012931797

CR Subject Classification (1998): C.2, D.4.4, H.3, H.4, D.2, K.6.5

LNCS Sublibrary: SL 5 – Computer Communication Networks and Telecommunications

Typesetting: Camera-ready by author, data conversion by Scientific Publishing Services, Chennai, India

Printed on acid-free paper

Springer is part of Springer Science+Business Media (www.springer.com)

Preface

The 4th International Workshop on Traffic Monitoring and Analysis (TMA 2012) was an initiative of the TMA COST Action IC0703 "Data Traffic Monitoring and Analysis: Theory, Techniques, Tools and Applications for the Future Networks" (http://www.tma-portal.eu/cost-tma-action).

The COST program is an intergovernmental framework for European cooperation in science and technology, promoting the coordination of nationally funded research on a European level. Each COST Action aims at reducing the fragmentation in research and opening the European research area to cooperation worldwide.

Traffic monitoring and analysis (TMA) is an important research topic within the field of computer networks. It involves many research groups worldwide that are collectively advancing our understanding of the ever-evolving Internet.

The goal of the TMA workshops is to open the COST Action research and discussions to the worldwide community of researchers working in this field. Following the success of the previous editions we decided to maintain the same format for this fourth edition: single-track full-day program. TMA 2012 was organized jointly with the 13th Passive and Active Measurement Conference (PAM 2012).

After a preliminary selection by the Workshop Chairs, 31 valid submissions were admitted to the review process, which finally resulted in the acceptance of ten full papers and eight short papers to the final program. Each paper was carefully reviewed by at least three members of the Technical Program Committee.

We address our sincere thanks to the authors for their submissions and to the Technical Program Committee members for their work during the reviewing process.

We hope you enjoy the proceedings.

March 2012

Antonio Pescapè
Luca Salgarelli
Xenofontas Dimitropoulos

Organization

Workshop Chairs

Antonio Pescapè — University of Naples, Federico II, Italy
Luca Salgarelli — University of Brescia, Italy
Xenofontas Dimitropoulos — ETH Zurich, Switzerland

Local Arrangements Chair

Christina Philippi — FTW, Austria

Website Maintainer

Pietro Marchetta — University of Naples, Federico II, Italy

Program Committee

Olivier Bonaventure — Université Catholique de Louvain, Belgium
Pere Barlet-Ros — Technical University of Catalunya, Spain
Kenjiro Cho — IIJ, Japan
Konstantina Papagiannaki — Telefonica, Spain
Alberto Dainotti — University of Naples Federico II, Italy
Alessio Botta — University of Naples Federico II, Italy
Nick Feamster — Georgia Tech, USA
K.C. Claffy — CAIDA, USA
Pietro Michiardi — Eurecom, France
Brian Trammell — ETH Zurich, Switzerland
Georgios Lioudakis — National Technical University of Athens
  (NTUA), Greece

Felipe Huici — NEC Labs Europe, Germany
Joseph Gasparakis — Intel, Ireland
Maurizio Dusi — NEC Labs Europe, Germany
Grenville Armitage — Swinburne University of Technology, Australia
Maria Papadopouli — University of Crete, KTH, FORTH, Greece
Gianluca Iannaccone — Intel Labs Berkeley, USA
Michela Meo — Politecnico di Torino, Italy
Marco Mellia — Politecnico di Torino, Italy
Steve Uhlig — Queen Mary University of London, UK
Yuval Shavitt — Tel Aviv University, Israel
Dario Rossi — TELECOM ParisTech, France
Fabio Ricciato — FTW and University of Salento, Italy

Table of Contents

Measurement for QoS, Security and Service Level Agreements

Tools for Network Measurement and Experimentation

Assessing the Real-World Dynamics of DNS

Andreas Berger and Eduard Natale

FTW Telecommunications Research Center Vienna
{berger,natale}@ftw.at

Abstract. The DNS infrastructure is a key component of the Internet and is thus used by a multitude of services, both legitimate and malicious. Recently, several works demonstrated that malicious DNS activity usually exhibits observable dynamics that may be exploited for detection and mitigation. Clearly, reliable differentiation requires legitimate activity to not show these dynamics. In this paper, we show that this is often not the case, and propose a set of DNS stability metrics that help to efficiently categorize the DNS activity of a diverse set of Internet sites.

1 Introduction

The Domain Name System (DNS) is a core component of the Internet. It is versatile enough to support any service from small web servers to large server farms with thousands of machines. It offers flexible decoupling of a service's domain name and the hosting IP address, which allows for load balancing strategies where one service name maps to multiple IP addresses over time. This flexibility is being abused by malicious services like botnets.

For increased resilience against malware mitigation, *Command-and-Control* (C&C) servers[1] continuously change their domain name and/or its IP resolution. These dynamics were recently targeted by several detection approaches (e.g., [1,3]). However, in contrast the DNS dynamics of legitimate services are largely unknown, and thus their impact on detection accuracy also is. Most prominently, Content Distribution Networks (CDNs) continuously change name-IP mappings to optimize host and network load. Similarly, virtual hosting providers operate large sites with thousands of servers, where each hosts more than one domain name. Dynamic DNS providers operate infrastructures for explicitly supporting continuous changes to name-IP mappings. And finally, large, growing services like Google, Youtube, and Facebook constantly increase and modify their existing installations, and thereby change DNS configurations.

In this paper we aim to quantify the extent of stability in DNS from the perspective of a single operator. Instead of focusing on malware detection, we rather show the dynamics of the system as a whole. We define a set of stability metrics and evaluate them using a large DNS trace. In particular, we show *what kind of stability* can be expected by *how many* and *which* Internet sites.

---

[1] Miscreants use C&C servers to control and monitor the activity of bots.

A. Pescapè, L. Salgarelli, and X. Dimitropoulos (Eds.): TMA 2012, LNCS 7189, pp. 1–14, 2012.

2   Background and Related Work

2.1   DNS Basics

The DNS infrastructure implements a distributed database with the main purpose of translating domain names to IP addresses. Specifically, DNS clients send a query (Q) for a domain name (e.g., `www.ftw.at`) to a recursive resolver (*RDNS*) which is usually run by the local network operator. By interfacing with the global DNS infrastructure, the resolver then attempts to find a matching address record (*A-Record*) for the queried domain name. The IP address included in the A-Record finally enables the client to contact the requested site.

Domain names are in the format `A.B.C.D`, where D is called the Top-Level Domain (TLD), C the 2nd-Level Domain (2LD), B the 3rd-Level Domain (3LD), etc. DNS implements domain name aliases, i.e. there can exist a sequence of mappings from a name to a *canonical name* (CNAME). In any case, a single name can map to multiple A-Records, that usually all map to the same redundant service. Each of these A-Records includes a Time-To-Live (TTL) value after which the RDNS must delete it from its local cache, and repeat the recursive fetching as soon as a new request for this domain arrives. Note that DNS clients are not forced to query the RDNS of their local network, but can use any publicly available one in the Internet.

2.2   Assumed Malware Model

In principle, malware uses the DNS infrastructure in the identical way as legitimate clients, with the only difference that the contacted servers host a malicious service, i.e. a C&C server. From that alone, it is impossible to reliably classify such DNS activity as malicious, due to the lack of any clear *signal* that is indicative for botnets. In other words, we simply cannot distinguish a stable botnet service from any other legitimate service.

However, due to the success of detection and mitigation strategies, botnets were forced to give up on this stability and implement a more dynamic DNS usage, so to improve on resilience and reliability of their infrastructure. Now, infected clients issue queries for periodically changing domain names. These queries resolve to a changing set of *IP addresses* of other infected clients, which serve as proxies to one or more C&C servers. This requires an *authoritative nameserver* that can be controlled by the botmaster, so to regularly update the name-IP mappings. In other words, malware changes either the mapping of a name to an IP (to avoid blocking of single IPs), or uses a new name for the same IP (to avoid blocking of names), or, in the worst case, does both. These types of malicious activity are commonly summarized as *Fast-Flux* (see Fig. 1).

2.3   Related Work

Most existing systems build on DNS data collected by a network operator. Bilge et al. recently proposed the EXPOSURE system for detecting malicious domains

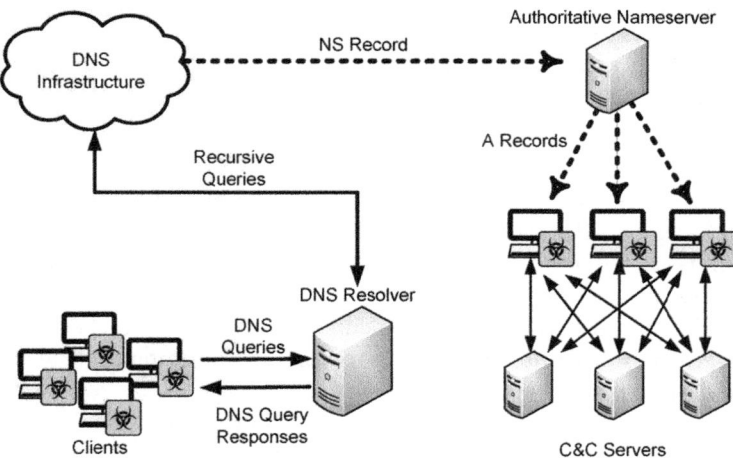

Fig. 1. Basic Malware Model

[3]. They use a total of 15 DNS features divided in four groups, namely time-based (i.e., the development of queries for a domain over time), DNS-answer-based (e.g., the number of A-records per query), TTL-based, and DNS-name based (i.e., the lexicographical structure of the queries). In particular they assume that short-lived domains are often malicious, and use a change point detection algorithm to discover domains that show popularity spikes which are followed by zero or only little activity. Using a similar approach, Antonakakis et al. propose their NOTOS system for building a dynamic reputation system for domain names [1]. They employ 41 features including BGP and AS information, as well as evidence collected from honeypots, to build a reputation database for domains. For training their classifier, they use traffic information about a manually categorized list of sites and retrieve different clusters like 'Popular Sites' := {Google, Facebook, . . . }. DNS activity that is later e.g. found to be closest to a 'bad' cluster, is consequently also classified 'bad'. It is therefore important to understand if (i) the categorization reflects similarity (e.g., are Google and Facebook really similar with respect to their DNS activity?), and (ii) if this activity is significantly different from malicious one.

A second branch of approaches bases on data collected at a domain's operator. Hao et al. study DNS lookup patterns by analyzing lookups to the TLDs .com and .net [5]. Thereby they concentrate on the distribution of queries from RDNSes, so to assess the cumulative DNS activity of entire networks. They find that malicious domains are more likely to be queried from a dynamic set of networks, and that many domains are either entirely legitimate or malicious. A related effort is described in [2]. Consequently, it seems reasonable to transfer trust from one domain name to a similar one. However, the required degree of similarity depends on the network, AS, or service operator in question. E.g., two domains ending on `google.com` are intuitively more suitable for transferring trust than two ending on `dyndns.org`.

A general problem of DNS analysis systems is that it is hard to evaluate the number of false positives. Usually, publicly available blacklists (e.g., `www.malwaredomainlist.org`) are used for result verification and algorithm training. However, these lists include domain names from a variety of sources, and the domain's activity must therefore not necessarily be related to dynamic DNS usage. Therefore, although a malicious domain is detected by a system, it is unclear if it was found because there is any observable dynamics in its activity, or just because the classifier was wrongly trained. Conversely, Alexa's list of popular domains[2] is often used for training machine learning algorithms or whitelisting. However, as only 2LD domain names are listed, the granularity at which sites are differentiated is rather low.

Partially similar to our ideas, Hu et al. analyze the global IP usage patterns of fast-flux botnets by conducting measurements from 240 geographically distributed network locations [6]. They mainly aim at differentiating malicious 'fluxy' activity from legitimate CDNs. Therefore, they continuously query a set of 5.169 suspicious domains to see how the returned DNS mappings develop over time. Most notably, they find that legitimate CDNs are trying to mostly return an IP address that is geographically close to the querying client, addresses involved in malicious activity are much more uniformly distributed over the world.

3   Methodology

Real-World Complications. In general, existing approaches strive to detect malicious activity by finding certain dynamics in DNS traffic, that can be differentiated from legitimate traffic. We investigate the same from the opposite direction: our aim is to find out to what degree *legitimate* DNS activity is *stable*. We conduct an extensive set of experiments on a DNS trace from a large operator network with several 100.000 customers. Over two weeks, we extracted and aggregated information from DNS 'NOERROR' responses, i.e. successful DNS query responses that provided an IP resolution for the requested domain name and thereby a name-IP mapping as described above. We used aggregation windows of two hours length and stored information about both the queried domain names as well as about the hosting ASes.

Google. Google's main search site is reachable under the address `www.google.{SUFFIX}`, where *SUFFIX* can be *com* or almost any country code (e.g., *de, fr, it, . . .*). However, all of these names are a CNAME alias of `www.l.google.com`[3]. This CNAME then finally resolves to a number of IP addresses, usually between one and six. In a 2-hour time window, we find 71 different IP addresses being used, which are from 17 different /24 networks (resp. from five /16 networks). One day later, only 14 new IP addresses are used in addition. This indicates

---

[2] http://www.alexa.com/

[3] Sometimes, also `www{X}.l.google.com` is used, where X is in [2,3,4,5].

that it might be possible to automatically learn the set of IP addresses used for a service, given enough time. But Google search is not the only service hosted at these IP addresses. Within the same two hours, Google's image search site images.1.google.com use 31 of the *same* IP addresses, plus two different ones in addition. A Fast-Flux detection approach must be able to cope with these 'natural dynamics', so not to wrongly classify these services as malicious, just because a large pool of IPs is used for many different domain names.

Another interesting detail is that the site www.youtube.com (which is also owned by Google) maps to the CNAME youtube-ui.1.google.com, thereby following the same naming scheme as shown above. However, it resolves to 44 IPs that are not overlapping with the other IP pool identified before, and usually there are only between one and four IPs advertised per query. Still, *all* of the IPs belong to the AS 'Google Inc.', i.e. Google seems not to host these services at some third-party hosting provider. In this case, knowing that a set of similar queries (i.e., *.1.google.com) is exclusively hosted in a single AS, could support DNS traffic classification.

Finally, a reverse lookup of the DNS name of any of the Google IPs always returns a DNS name of the format {X}.1e100.net (e.g., fx-in-f147.1e100. net). Apparently, Google implemented a common naming scheme for all their services, which might help to identify some site as belonging to this family of services[4].

Facebook. Within two hours, we find 54.691 different domain names with the format {X}.{Y}.channel.facebook.com, where X and Y are always numbers (e.g., 01371354742.67.channel.facebook.com). However, they map to only 28 IPs in three different /24 networks. One day later, we observe 80.416 new domain names with the same format, still mapping to the same 28 addresses. Facebook seems here to 'abuse' the DNS system for short-lived, extremely dynamic service identification, using some form of wildcard name-IP matching (i.e., *.channel. facebook.com). While it would be easy to whitelist this particular phenomenon, a universal detection approach for malicious domains must be designed such that these short-lived domains are recognized as normal, legitimate activity.

Amazon. Amazon shows a two-fold face: on the one hand, their main site www.amazon.com maps to only three different IPs in two hours, all from their own AS 'Amazon.com, Inc.'. In strong contrast to Google, their European branches (*.de, .fr, .it, ...*) map to a small set of IP addresses in the AS 'Amazon EU DC AS'. This activity is therefore highly stable. However, for Amazon's cloud services, the observed activity is completely different. Within two hours, we observe 5.830 queries of the format {X}.profile.{Y}.cloudfront.net (e.g., a5e0129bd6b7bd4ef1e4f83b979a1216e.profile.dub2.cloudfront.net)   on 3.903 different IPs. Each name usually maps to 8 different IPs, all of them belonging to either AS 'Amazon.com Tech Telecom', 'Amazon EU DC AS', or 'Amazon.com, Inc.'.

---

[4] 1e100, or $1 \cdot 10^{100}$, is called a 'Googol' by mathematicians.

Akamai. Akamai operates one of the largest Content Distribution Networks (CDNs), a large service diversity is therefore to be expected. Within two hours, we observe 3.168 CNAME aliases with the format a{X}.{Y}.akamai.net, where X is a number (roughly in [1,2.000]), and Y is a short string (e.g., a1254.w7. akamai.net). In total, these CNAMES map to 1.729 different IP addresses. Additionally we find 20 CNAMES with the similar format a{X}.{Y}.akamai.net. 0.1.cn.akamaitech.net. While almost all of these IPs belong to the AS 'Akamai Technologies European AS', we find 31 other ASes of backbone operators and Internet providers. Simply counting the number of different ASes to which a domain name is associated with over time and expecting to find just malicious activity over a certain threshold, is therefore misleading.

In addition, for understanding the complications of these configurations, consider the following example: The domain name js2.wlxrs.com was found to be hosted in the AS 'Akamai Technologies European AS', specifically on a set of servers that follow the naming scheme introduced above. It is a CNAME alias of login.live.com, the login site of Microsoft's *Live* service. Within the same two hour time window however, this site maps much more often to the CNAME alias login.live.com.nsatc.net, and points to an IP in AS 'Microsoft Corp'. In summary, we have to deal here with a popular service that maps to multiple CNAME aliases and points to different IPs in different networks. This activity is prone to be misclassified as Fast-Flux, and poses a significant challenge to botnet detection approaches.

No-IP.org. As a final example, we consider the popular dynamic DNS provider No-IP.org, which operates authoritative nameservers for a number of domains. Sub-domains of these can be registered by anybody at no cost. Within two hours, we observe 109 queries of the format {X}.no-ip.org that resolve to 74 unique IPs in 52 different ASes. This corresponds to 69 different /24 networks, and resp. 61 different /16 networks. Clearly, such domains can be easily taken for malicious ones when looking only at their distribution over the networked world. Given that such free, disposable domains names seem like a perfect way for a miscreant to host a C&C server, botnet detection approaches need to find a way to distinguish such legitimate dynamic DNS usage from malicious one.

3.1 Defining DNS Stability

Many of the given examples exhibit different, dynamic DNS features at the first glance, although they represent highly stable, legitimate services. Thus, DNS stability means different things for different services. In the following we define a flexible metric to qualitatively assess the degree of stability of a particular service. We use the following notation: $\mathcal{Q}$ and $\mathcal{P}$ are the sets of domain names and IP addresses respectively. We define $\mathcal{M}_\mathcal{Q}$ as the set of tuples $\langle q \in \mathcal{Q}, \mathcal{P}^* \rangle$, where $\mathcal{P}^* \subseteq \mathcal{P}$ is defined by $\{p \in \mathcal{P} \ s.t. \ q \ resolves \ to \ p\}$. Similarly, $\mathcal{M}_\mathcal{P}$ is the set of tuples $\langle \mathcal{Q}^*, p \in \mathcal{P} \rangle$, where $\mathcal{Q}^* \subseteq \mathcal{Q}$ is defined by $\{q \in \mathcal{Q} \ s.t. \ q \ resolves \ to \ p\}$.

The simplest form of a *stable, unidirectional mapping* is then given when for a tuple $\langle q, \mathcal{P}^* \rangle$ there exists only a single element in $\mathcal{P}^*$ (and analogously for tuples $\langle \mathcal{Q}^*, p \rangle$). Therefore, a *stable, bidirectional mapping* is given when any q maps to a single p, and p exclusively maps to this q. For simplicity, we refer to this as **1:1** stability in the following. The set $\mathcal{S}$ of 1:1 stable mappings is thus defined as $\mathcal{S} := \mathcal{M}_{\mathcal{Q}}^s \cap \mathcal{M}_{\mathcal{P}}^s$, where $\mathcal{M}_{\mathcal{Q}}^s \subseteq \mathcal{M}_{\mathcal{Q}} := \{\langle \{q\}, \mathcal{P}^* \rangle \ s.t. \ |\mathcal{P}^*| = 1\}$ and $\mathcal{M}_{\mathcal{P}}^s \subseteq \mathcal{M}_{\mathcal{P}} := \{\langle \mathcal{Q}^*, \{p\} \rangle \ s.t. \ |\mathcal{Q}^*| = 1\}$.

For clarity, consider the following real-world example of a 1:1 mapping: the queried name `www.ftw.at` points *always* to the IP address `213.235.244.145`, and no other query than `www.ftw.at` *ever* points to `213.235.244.145`. These types of mappings are extremely common, and should clearly never be reported by a malware detection approach.

In order to address more complex types of mapping stability, we define the following relaxations: $k - LD(query)$ specifies the k-level domain name of $query$, e.g. 2-LD(`www.ftw.at`) is `ftw.at`. Similarly, $j - NW(IP)$ gives the $/j$-network of an IP address, e.g. 24-NW(`213.235.244.145`) is `213.235.244.0/24`. This is motivated by the fact that both the name and the address can vary at different degrees, while still representing the same service. I.e., large Internet sites assign a network segment to single services instead of a single IP and, on the other hand, often more than one name points to a single IP address. And as we can see clearly with e.g. the various Google services, also both name and address vary often at the same time. Still, they do not vary arbitrarily: multiple IP addresses for a service are often in the same network segment, and multiple names frequently share a common suffix.

The set of stable mappings $\mathcal{S}$ is still derived as shown above. The names and addresses in the input sets $\mathcal{Q}$ and $\mathcal{P}$ are now the results of applying k-LD(query) and j-NW(IP). For simplicity, we refer to $k - LD : j - NW$ mappings as $k : j$, e.g. a 2-LD:24-NW mapping is abbreviated as 2:24. As an additional convention, and in line with the 1:1 mapping defined above, we use *1* for describing the case when we do not apply k-LD or j-NW at all[5]. I.e., a 1:24 mapping describes a single domain name that is exclusively hosted by an $/24$ network. Conversely, a 3:1 mapping describes all services that share the same 3LD name and are exclusively hosted at the same IP address. Also note that 1:32, ∞:1, and ∞:32 are equivalent to 1:1.

The introduced bidirectional definitions are strong in the sense that they allow for only little variation of a domain's name and the IP addresses hosting them. Conversely, everything that is considered stable according to any of these definitions exhibits extremely little dynamics, and does therefore not match the malware model introduced in Section 2.2. However, as our analysis of legitimate sites above shows, also many legitimate sites are not considered stable according to these definitions. Many of these sites share a common reason for that, namely the fact that they use a set of IP addresses from *different* networks for one or more domain names. Identifying a network mask that matches all these mappings *but no others* is therefore in general impossible. However, we can exploit the stability of the geographical position of these IPs, as many different ranges of

---

[5] For practical reasons, it makes little sense to consider 1LD (i.e., TLD) names or $/1$ networks (i.e., half the Internet) anyhow, so we accept this contradiction here.

Algorithm 1. Find-k:j-mappings

---

input : DNS-Records, k, j
output: stables, bfd, bfa

stables ← Set(); bfd ← BloomFilter(); bfa ← BloomFilter();
foreach *record* r *in* DNS-Records **do**
 dname ← k-LD(r.*name*, k);
 addrs ← Get_IPs(r.*A-Records*);
 if dname *in* bfd **then** bfa.rel_add(addrs);continue;
 foreach *a in* addrs **do**
  ip ← j-NW(a.*ip*, j);
  if ip *in* bfa **then** bfa.add(addrs); bfd.rel_add(dname);break;
  if dname *not in* stables *and* ip *not in* stables **then**
   stables.add(dname,ip); continue;
 end
 if stables[dname]=ip *and* stables[ip]=dname **then**
  continue;
 end
 if stables[dname]≠ ip **then** bfa.add(addrs); bfd.rel_add(dname); break;
 if stables[ip]≠ dname **then** bfd.add(dname); bfa.rel_add(addrs); break;
 end
end

---

a single organization are often registered at the same location. We define this additional type of stability as $1 \to GEO$ stability, i.e. a domain name is considered stable when it always resolves to an IP address at the same longitude/latitude.

3.2   Implementation

The basic idea is to find the set of stable mappings in a DNS trace by testing each mapping against each definition of stability as defined in the previous section. As the processing of the trace proceeds, new stable mappings are added and those contradicting the particular definition of stability are removed. Once a k-LD name or a j-NW network has been removed from the stable set, no other mapping that contains either of them can be considered stable. We implement this by using two Bloom Filters [4] for efficiently storing the already removed domain names (*bfd*) and addresses (*bfa*). The entire procedure is shown in Algorithm 1. Note that BloomFilter.add(x) also removes x from the set of stable mappings. BloomFilter.rel_add(x) additionally adds all *related* entries, i.e. those mappings that were stored previously. For example, if *stables* contains the mapping www.example.com:1.2.3.4 and a DNS record with www.example.com:5.6.7.8 is observed, then www.example.com, 1.2.3.4, and 5.6.7.8 are removed from *stables* and are added to the Bloom Filters.

As Bloom Filters are probabilistic data structures, false positives may be reported. That is, the filter wrongly reports a certain element to be stored, while it is not. The false positive rate is a function of the filter size, the number of hash functions used, and the number of inserted elements. Knowing the total number of unique domain names in our trace (1.444.735), and therefore the maximum required filter size, we set the number of hash functions for *bfd* to 34 to guarantee a false positive rate of $\leq 0.01\%$. Using again 34 hash functions, we set the filter size of *bfa* to 700.000 to achieve the same false positive rate. During the experiments, the maximum number of IP addresses stored in one instance of *bfa* was 313.863 (for experiment 2:1). The filters are stored after each run of the experiment, so that we can stack them together in different ways for quickly getting different views on the data, that depend on the order of the individual experiments. However, note that due to the non-zero false positive rate we occasionally misclassify single mappings.

For the 1→GEO experiment we slightly modified Algorithm 1 to consider ⟨*longitude, latitude*⟩ tuples instead of IP addresses. We used MaxMind's free version of their GeoIP City database[6] for retrieving the geographical location of IP addresses.

4   Results

We analyzed the aforementioned DNS trace which was created by capturing and parsing all packets on UDP port 53 in a large European Internet provider's network. The trace includes all DNS activity from 2010-11-21, 21h to 2010-12-5, 21h (i.e., 14 days). We extract from this data the DNS 'NOERROR' query responses and aggregate them on two-hour time bins using the queried name as a key. That is, for every two hours we output a list of queried domain names together with the number of queries they received and the list of IP addresses in the responses' A-Records. During processing, we then remove those records that received less than three queries. Furthermore, we remove the prefix www. from all queries, and therefore treat www.X.com the same as X.com[7]. In total, taking into account these measures, we observe 1.444.872 unique queried names which are hosted in 21.968 different autonomous systems and for which we received 1.025.858.834 individual DNS queries.

4.1   1:1 Stability

We found 310.788 stable 1:1 mappings in our trace, which represent 21% of the unique queried names in the entire trace. In other words, one fifth of the sites host only a single service on a single IP address. Most likely, these are small companies or private websites, that do neither require redundant hosting nor multiple services on the same IP. However, this set is constantly evolving. Figure 2 shows the innovation over time of the number of stable mappings. For better visibility, we also provide a zoom-in on the tail of the results (from 2010-11-29 to the end of the trace). After the first day of analysis, the number of newly found stable mappings decreases as popular 1:1 stable sites are already known by then. However, it is interesting to note that even at the end of the trace, i.e. after

---

[6] http://www.maxmind.com/app/city

[7] Many sites use the same mapping for both, as users often do not use the www.-prefix.

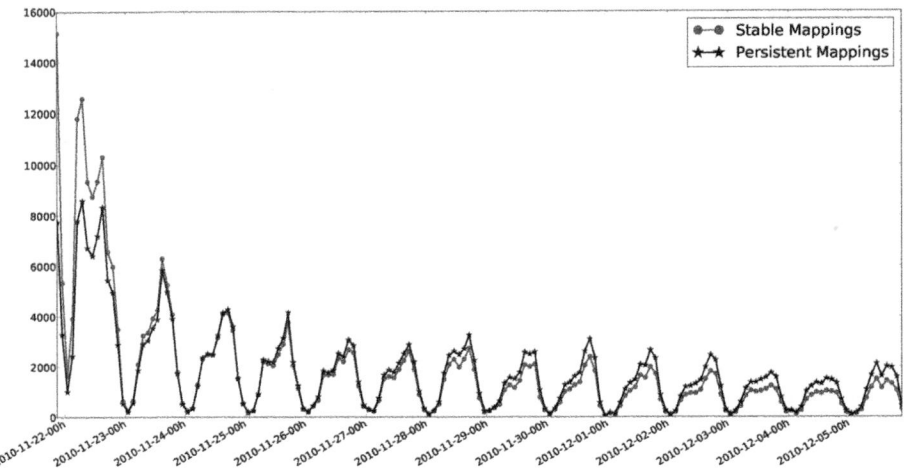

Fig. 2. 1:1 mappings: innovation over time

almost two weeks, the set of stable mappings grows at least by a few hundred in every time bin. Note that this set could theoretically also shrink, given there would be more mappings removed than new ones are introduced, which is not the case. In other words, there are always more entirely new mappings, than new name or address conflicts (due to reuse) being found.

Of course, it is possible that a new mapping is considered stable in one time bin, but gets removed again in a subsequent one. Therefore, we assessed the number of *persistent mappings* at each time bin t, i.e. those that are first seen at time t, but are still considered 1:1 stable at the end of the experiment. Interestingly, the number of new, persistent mappings decreases rather slowly. This suggests that a system for detecting malicious domains should be designed such that it can adapt to these dynamics, i.e. purge historic, inactive mappings to avoid future name-address conflicts, and learn at the same time about new ones. For many algorithms that means that continuous retraining should be foreseen.

4.2   k:j Stability

After having identified the set of 1:1 stable mappings, we continue by gradually relaxing our definition of stability. We define three separate experiments: first we analyze the special cases 1:j and k:1. Then, we investigate several other combinations of k:j mappings (i.e., with k,j $\neq$ 1).

1:j. This experiment targets domain names that *exclusively* use a set of IP addresses instead of a single one. We define the set of netmasks as $\mathcal{J} = \{31, 30, 29, 28, 27, 26, 25, 24\}$, and run one experiment for each $j \in \mathcal{J}$. We start with the largest value (31) and continue with 30, 29, etc. By addressing more and

Table 1. 1:j results

j	||stables||	|bfd|	|bfa|	I	j	||stables||	|bfd|	|bfa|	I
31	263.945	1.180.927	205.031	715	**27**	123.756	1.321.116	144.004	182
30	227.344,	1.217.528	189.549	496	**26**	96.629	1.348.243	127.371	112
29	190.168	1.254.704	176.142	356	**25**	73.996	1.370.876	110.793	58
28	155.741	1.289.131	160.276	295	**24**	55.689	1.389.183	94.504	45

more addresses, our definition of stability gradually 'widens'. Note that, as soon as it gets too wide so that other services 'pollute' the IP range, 1:j stability is not given anymore. Table 1 shows the results, adopting the terminology from Algorithm 1. The number of stable mappings decreases from one experiment to the next, due to overlaps in network ranges from different services. Conversely, the number of elements in *bfd* steadily increases, due to more and more address conflicts. The elements stored in *bfa* are in this case /j networks, hence the number of blocked IP addresses also steadily increases. In addition, we compute the innovation I from one step to the next. As a first reference we use the results from the 1:1 experiment, and find that 1:31 matched 715 new domain names. 1:30 finds then 496 as compared to 1:1∪1:31 etc. Note that the found domain names use the individual IP ranges *exclusively*, which represent in total 42.558 IP addresses. We find no dominant set of sites with 1:j stability, but many of them are larger services like `europe.battle.net`, an online gaming site, which is found to be 1:24 stable.

k:1. Similarly, we define the set of domain levels to analyze as $\mathcal{K} = \{6, 5, 4, 3, 2\}$, so to target domain names that share are common suffix but are all *exclusively* hosted at the same IP, and run one experiment for each $k \in \mathcal{K}$. Due to lack of space, we report here only the innovation from experiment to experiment. As before, we use the set of 1:1 mappings as a first reference, and compute the innovation in the order of decreasing k. The results are $\{3, 6, 296, 563, 2.526\}$, i.e. in total we find 3.394 k:1 stable mappings. Note that the number of mappings is not equal the number of matched domains here, as k-LD represents a not-unique domain pattern. As it is considered new compared to 1:1, there must at least be two domains matching each. The total domain counts matched per $k-LD$ pattern are $\{9, 19, 10.267, 3.386, 8.257\}$. The high count of 10.267 domains is almost exclusively due to a large number of `{X}.webim{Y}.webim.myspace.com`. Furthermore, we find 643 domain names `{X}.jim{Y}.mail.ru` that are 3:1 stable.

k:j. Finally, we run a set of 40 experiments using all combinations for $k \in \mathcal{K}, j \in \mathcal{J}$. The individual innovation is shown in Figure 3. On the right, the number of new k-LD patterns is shown (e.g, `*.example.com`), while on the left we show the number of domains matching the individual patterns. We compute innovation in a similar way as before, using as a first reference the set of stable mappings found by all previous experiments. Following our idea of gradually widening the

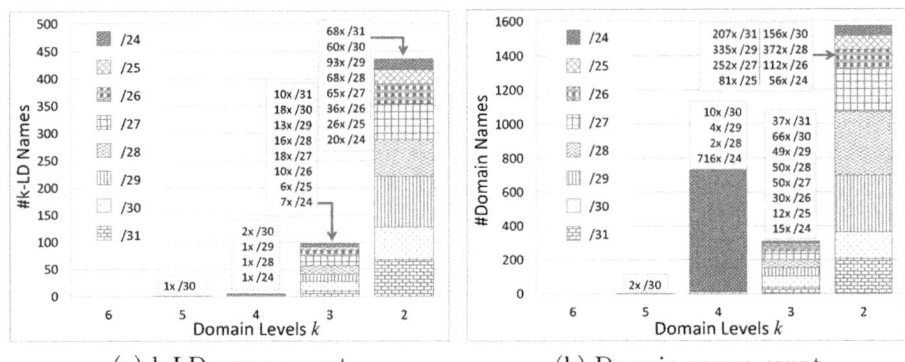

(a) k-LD names count   (b) Domain names count

Fig. 3. k:j mappings: innovation per k,j configuration

definition of stability, we proceed in the following order: 6:31→ 6:30 → ... 5:31 →
... 2:24. In general, the set of found domains is rather diverse, with no dominant
service. The largest share of domains is in 4:24 and is exclusively caused by
`*.profile.sin2.cloudfront.net`.

4.3 1→GEO

In total, we find 1.391.251 1→GEO mappings, i.e. 96% of the unique queries in
our trace. Out of that, 1.053.657 were not already found by the k:j experiments.
Among those are 291.977 unique `*.channel.facebook.com` and 9.402 unique
`*.cloudfront.net`. The only two similar domain names *not* in this set are `0.53.`
`channel.facebook.com`, which appear in two different ASes (US, Ireland), and
the 716 domains `*.profile.sin2.cloudfront.net` which were already found
previously by the 4:24 experiment.

Although it is quite surprising that 1→GEO stability is given for such a large
number of mappings, this is clearly a rather coarse measure compared to the
other definitions of stability. In particular, note that 1→GEO stability is not
sensitive to IP address reuse, i.e. assigning multiple domain names to the same
IP address. In order to get an idea of how many legitimate sites are among these
mappings, we tried to resolve their domain names again. Given that since the
recording of the DNS trace almost one year passed, we reason that any domain
name that would still be resolvable to an IP address, is most likely legitimate[8].
We randomly selected 105.365 (i.e., 10%) domain names from the set of mappings
that were exclusively found by 1→ GEO and found that 103.464 still resolved
to an IP address. 1.901 (i.e., 1.8%) domains were not resolvable anymore. We
conclude that the majority of 1→GEO domains is therefore legitimate.

---

[8] Note that due to DNS domain parking we might wrongly assume that a site is still
active, as we would be redirected to a default site. The majority of sites does not
implement this though, so we consider this bias acceptable.

Finally, we build a set of stable mappings from the individual results of all experiments which contains 1.391.256 entries, i.e. 96% of all unique queried names are considered stable. Interestingly, these correspond to only 492.730.776 out of the 1.025.858.834 individual queries in our trace, i.e. 48%.

4.4   Analyzing the Remaining Domain Names

We found 53.616 domain names in 3.837 ASes that were not considered stable according to any of our definitions of stability. The two most popular sites `www.google.com` and `www.facebook.com` alone received 45.941.330 and 23.813.897 queries respectively. The top-{10,30,50,100} domain names received {154.129.864, 224.851.245, 320.273.845, 373.019.662} queries. The top-10 ASes hosting these 53.616 domains were Microsoft Corp, Peer 1 Network Inc., Layered Technologies, Inc. (two separate ASes), Akamai Technologies European AS, 1&1 Internet AG, Internet Systems Consortium, Inc., Georgia Institute of Technology, MX Logic, Inc., and Deutsche Telekom AG. In total, the top-10 ASes hosted one or more IP addresses of 54% of the found domains, which attracted 57% of all individual queries.

Among these highly dynamic domains it was easy to spot 2.565 ones (partially) hosted in Georgia Tech's AS, as they stood out due to their random-looking domain names. Apparently they belong to a botnet that was caught by Georgia Tech's honeynet research project[9].

5   Discussion and Conclusion

Due to the multitude of different and evolving services, the DNS is constantly changing. New domain names and new IP addresses appear and disappear on a daily basis. The specifics of this activity depend on the type of operator that hosts a given service.

The k:j mappings represent a significant share of the set of unique queries (22%), and a malware detection approach could therefore significantly reduce its workload by keeping track of these. Instead of analyzing these queries over and over again, a detection system could then be able to rule them out quickly. However, as the large set of 1:1 stables is constantly changing, the system must remove inactive mappings after some time[10].

1→GEO stability provides a simple way for discovering services that are not entirely stable, but are at least not hosted in multiple, geographically distributed networks. However, IP reuse cannot be detected. Finally, the relatively few sites remaining are highly dynamic and attract the lion's share of the traffic. Differentiating the legitimate ones from the malicious ones is hard not only because of the similar DNS activity patterns, but also because of the large amounts of data to process.

---

[9] `http://users.ece.gatech.edu/owen/Research/HoneyNet/HoneyNet_home.htm`

[10] See also Hao et al. [5], who found 5.711.602 new {.com, .net} domains in only 2 months.

We think that these three different classes of DNS activity must be treated separately. k:j stable mappings are highly likely to be legitimate, and should be removed from any further analysis as soon as possible. 1→GEO mappings deserve further analysis of the IP addresses being used, to detect reuse. For the remaining sites, it is definitely worth to whitelist single domain names, as already the top-10 would decrease the number of queries to process by 15%. However, it seems risky to whitelist entire domains or ASes, as the gain in terms of queries that do not have to be analyzed is considerably less, but the risk of whitelisting non-legitimate sites in that domain (that are somehow user-controlled, as e.g. cloud services or personal homepages) is not controllable.

Acknowledgements. This work has been supported by the Austrian Government and by the City of Vienna within the competence center program COMET. The research leading to these results has received funding from the European Union's Seventh Framework Programme [FP7/2007-2013] under grant agreement n° 257315. The work was partially supported by the project DARWIN3. In particular, we would like to thank our colleague Arian Bär for supporting us with the DNS data extraction.

References

1. Antonakakis, M., Perdisci, R., Dagon, D., Lee, W., Feamster, N.: Building a dynamic reputation system for DNS. In: Proc. of USENIX Security, Berkeley, CA (2010)
2. Antonakakis, M., Perdisci, R., Lee, W., Vasiloglou, N., Dagon, D.: Detecting malware domains at the upper DNS hierarchy. In: USENIX Security, Berkeley, CA (2011)
3. Bilge, L., Kirda, E., Kruegel, C., Balduzzi, M.: EXPOSURE: Finding malicious domains using passive DNS analysis. In: Proc. of NDSS, San Diego, CA (2011)
4. Bloom, B.H.: Space/Time trade-offs in hash coding with allowable errors. Communications of the ACM 13, 422–426 (1970)
5. Hao, S., Feamster, N., Pandrangi, R.: An internet wide view into DNS lookup patterns. Tech. rep., VeriSign Incorporated (June 2010)
6. Hu, X., Knysz, M., Shin, K.G.: Measurement and analysis of global IP-usage patterns of fast-flux botnets. In: Proc. of INFOCOM, Shanghai, China, pp. 2633–2641 (April 2011)

Uncovering the Big Players of the Web

Vinicius Gehlen[1], Alessandro Finamore[2],
Marco Mellia[2], and Maurizio M. Munafò[2]

[1] CNIT - UoR c/o Politecnico di Torino
[2] Dept. of Electronics and Telecommunications - Politecnico di Torino
lastname@tlc.polito.it

Abstract. In this paper we aim at observing how today the Internet large organizations deliver web content to end users. Using one-week long data sets collected at three vantage points aggregating more than 30,000 Internet customers, we characterize the offered services precisely quantifying and comparing the performance of different players. Results show that today 65% of the web traffic is handled by the top 10 organizations. We observe that, while all of them serve the same type of content, different server architectures have been adopted considering load balancing schemes, servers number and location: some organizations handle thousands of servers with the closest being few milliseconds far away from the end user, while others manage few data centers. Despite this, the performance of bulk transfer rate offered to end users are typically good, but impairment can arise when content is not readily available at the server and has to be retrieved from the CDN back-end.

1 Introduction

Since the early days the Internet has continuously evolved. With the aggregation of different technologies, the convergence of telecommunication services and the birth of innovative services, nowadays understanding the Internet is very complicated. Measurements are thus becoming more and more a key tool to study and monitor what the users do on the Internet and how can the Internet meet users demand.

The last years witnessed the convergence of services over the web thanks to the increasing popularity of Social Networks, File Hosting, Video Streaming services. All of them involve a large amount of content that has to be delivered to a humongous number of users. Delivering such impressive volume of content in an efficient and scalable way has become a key point for the Internet service and content providers. Content Delivery Network (CDN) and data centers have become the vital elements to support such growth. It is indeed not surprising that 40% of the inter-domain traffic is generated by few *organizations*[1], with Google leading the group, and well-known CDN players closely following [1]. Companies

---

[1] In this paper we use the generic term "organization" to state commercial companies that manages web servers, caches, CDNs, and offering services directly to end users or from third parties.

A. Pescapè, L. Salgarelli, and X. Dimitropoulos (Eds.): TMA 2012, LNCS 7189, pp. 15–28, 2012.
© Springer-Verlag Berlin Heidelberg 2012

like Akamai, Limelight and Level3, which have been deployed in the last decade, are now crucial for the majority of the web-based services. Several studies in the literature have focused on CDN architectures, trying to reverse engineer the server placement and load balancing algorithms [2–5]. These works usually focus on big players as Akamai [3, 4] or, more recently, YouTube [5]. However, besides CDN companies there are many others smaller organizations which provide web services as well [6], and overall their role is not marginal. Another common treat among previous works is the extensive usage of active experiments, i.e., abusing of synthetic traffic which does not reflect the real usage and users' habits, and thus does not allow to gauge the actual performance offered to the end users.

Our work goes in a different direction. We consider actual traffic originated from the activity of more than 30,000 residential customers monitored from three different vantage points of an Italian ISP. Analysing the traffic they generated for one week, we aim at identifying i) which services they access, and ii) how organizations handle it. We neither focus on single organization, nor service. We instead consistently compare and quantify the traffic users exchange with server managed by top players handling web traffic by looking for both what kind of service and performance they offer, and how dissimilar their architecture looks like.

It is important to highlight that we are not aiming at a complete characterization of all the possible services the big organizations provide. We instead would like to start understanding what type of traffic they have to handle, how do they handle it and what kind of performance they offer. Moreover, given we are considering only one ISP, we take a picture from the same point of view of the ISP customers. Clearly, we expect differences when observing from other ISPs.

In the following we report some of the key findings:

- the top 10 organizations handle 65% of the total web traffic. This is invariant respect to the location of the users, the time of the day, and the day of the week;
- nearly all of the top 10 organizations provide the same type of content, ranging from video services, to software download, to hosting generic content;
- only Google, Facebook and Akamai handle service over HTTP/SSL. Encryption is rapidly increasing its share due to users enabling security features in social network systems;
- the organizations adopt different architectures, with more than 10,000 IP addresses managed by Akamai but only 338 by Facebook according to our data sets;
- some organizations serve traffic within few milliseconds far away from ISP customers, while other organizations manage few data centers that are placed in other countries or even continent;
- traffic balanced among servers is the same over time for all organizations except for Google which instead routes traffic to different data centers at different time of the day;
- bulk download rate offered by different organizations is good, with Limelight taking the lead. Only Akamai shows some poor performance when the requested content is not readily available at the cache.

Table 1. Data sets used in this paper

Name	Volume [GB]	Flow [M]	# Servers	# Clients
VP1	1745 (35%)	16 (63%)	77,000 (0.14%)	1534 (99%)
VP2	10802 (44%)	84 (53%)	171,000 (0.6%)	11742 (97%)
VP3	13761 (35%)	125 (52%)	215,000 (0.5%)	17168 (98%)

In summary, our work aims at unveiling some aspects of modern web browsing. We believe this information is useful for both the service providers and the research community to shed some light on the role big organizations have on the current Internet.

2   Methodology and Data Sets

The data sets has been collected using Tstat [7], the Open Source packet sniffer developed in the last 10 years by the Telecommunication Network Group (TNG) at the Politecnico di Torino. Tstat rebuilds TCP connections by monitoring traffic sent and received by hosts. The connections are then monitored as to provide several types of Layer-4 statistics. Using Deep Packet Inspection (DPI) and statistical techniques [8, 9], each connection is further classified based on which application has generated it.

We collected data sets from three Point of Presence (PoP) aggregating thousands of customers of an Italian ISP. At each vantage point (VP), we installed a probe consisting of an high-end PC running Tstat and monitoring all the traffic sent and received by hosts inside the PoP. This paper focuses on one-week long data sets collected simultaneously at the three vantage points starting from 12:00 AM of June 20th, 2011. Each data set is composed of TCP flow-level logs. Each line logs the statistics of a specific TCP flow while the columns report specific features[2].

In this paper we analyze organizations handling content related to web-based services. Thus, we focus only on HTTP (and HTTPS) traffic since the remaining part of the traffic is based on applications and protocols which usually are not handled by these organization. The commercial MaxMind [10] "Organization" database has been used to associate each server IP address with the organization which owns it.

Tab. 1 summarizes the data sets reporting the name used in the remaining of the paper, the volume due to HTTP traffic in terms of bytes and flows, the number of distinct HTTP servers contacted by clients and the number of distinct internal hosts that generated HTTP traffic during the whole week. In brackets we report the corresponding fraction with respect to the total traffic. For example, consider VP1 data set. It contains 1.7 TB of HTTP traffic carried in 16 millions of TCP flows. It corresponds to 35% of the total volume exchanged

---

[2] A description of all statistics is available from
http://tstat.tlc.polito.it/measure.shtml

Table 2. Characterization of the Top 10 organizations

Organization	Volumes			Most known services		
	%B	%F	%Clients	Video Content	SW Update	Adv. & Others
Google	22.7	12.7	97.1	YouTube	-	Google services
Akamai	12.3	16.7	97.2	Vimeo	Microsoft, Apple	Facebook static content, eBay
Leaseweb	6.3	1.1	64.3	Megavideo	Mozilla	publicbt.com
Megaupload	5.5	0.2	15.6	Megavideo	-	File hosting
Level3	4.7	1.9	79.7	YouPorn	-	quantserve, tinypic, photobucket
Limelight	3.9	1.6	72.5	Pornhub, Veoh	Avast	betclick, wdig, trafficjunky
PSINet	3.2	0.2	44.6	Megavideo	Kaspersky	Imageshack
Webzilla	2.9	0.3	13.2	Adult Video	-	filesonic, depositfiles
Choopa	1.5	0.01	5.7	-	-	zShare
OVH	1.0	0.7	63.1	Auditude	-	Telaxo, m2cai
Facebook	0.9	4.2	90.6	Facebook	-	Facebook dynamic content
total	64.9	39.6	-			

by hosts in that vantage point, i.e., 63% of TCP flows. The remaining part of the traffic is due to other applications like email, chat, and most of all, peer-to-peer (P2P) applications. 77000 distinct HTTP servers have been contacted by 1534 distinct customers hosts. HTTP servers are a small fraction of contacted hosts (the majority of them are due to P2P applications), but almost any PoP internal customer generated HTTP traffic. VP2 and VP3 data sets aggregate a larger number of local clients which is reflected in terms of volumes and number of contacted servers.

3   Traffic Characterization by Organization

We now group HTTP traffic based on the organization that owns the server IP address. We start by answering simple questions like which organizations are responsible for the largest majority of traffic, which kind of content they handle, and how the traffic changes when considering different vantage points or time of the day.

3.1   Share and Popularity

Tab. 2 shows the top 10 organizations ranked with respect to the volume of HTTP traffic they account for. On the left, after each organization name, we report the fraction of bytes and flows the organization handled, and the fraction of customers that generated more than 100 flows with servers of that organization during the week. Results reported refer to VP2 data set and are similar on other vantage points. The rightmost columns report the kind of services offered by the organizations, identifying three coarse categories, namely Video content, Software update and Other services as advertisement or File Hosting.

Several considerations hold. Beside the expected presence of big players as Google or Akamai, there are less known organizations as Choopa or PSINet responsible for a non negligible fraction of traffic. Even if this result is specific for the operator we are considering, knowing which are the top organizations the customers exchange traffic with is a key information. Interestingly, the top ten organizations account for 65% of the HTTP traffic volume. Overall, we have found more than 10,000 organizations; yet, 90% of the volume is managed by 70 companies only. This testifies that in the nowadays Internet there are plenty of companies providing an high number of services but the majority of the traffic is generated by very few of them, as already noted in [1].

Considering the popularity of the organizations among ISP customers, we notice that some are handling niche services, like the one supported by Choopa or Webzilla, while others are practically used by most users. It is not surprising that 97% of users have contacted Google, Level3 or Akamai administered servers. Instead it is more surprising that 91% of users contacted one Facebook server or 63.1% used some service handled by OVH. Considering Facebook, most of its static content like pictures and videos, is provided by the Akamai CDN. Indeed, we have found that more than 73% of the bytewise volume related to Facebook is outsourced to Akamai and overall Facebook content corresponds to 15% of the total volume handled by Akamai. The Facebook dynamic content, including all embedded objects found in other web pages as the "Like" button, is directly handled by the Facebook servers. The pervasiveness of these objects on the web causes a high probability of connecting to a Facebook server, even without accessing to the Social Network site. For OVH we believe that the presence of several advertisement services that it hosts causes most of customers to contact the OVH servers.

For the remaining part of this work, we will focus only on the most important organization listen in Tab. 2, namely Google, Akamai, Leaseweb, Level3, Megaupload, Limelight and Facebook.

3.2   Spatial and Temporal Difference

We now investigate how the traffic changes considering i) different vantage points, ii) different time of day and iii) different days. Are those results invariant to the spatial and temporal choice of observation? Fig. 1 reports the daily evolution of the HTTP bitrate breakdown among the considered organizations for the VP2 data set. Each bar refers to a 2 hour time window. Fractions are derived by averaging the traffic in all days of the data set. As expected, traffic volume follows the typical day/night pattern. However, in each period of the day, the fraction of traffic handled by the different organization is very stable, so that practically in all period they handle 60% of the web traffic. Focusing on peak hour time from 22:00 to 00:00, Fig. 2 (left) compares the traffic share considering different days of the week. Also in this case, marginal changes are

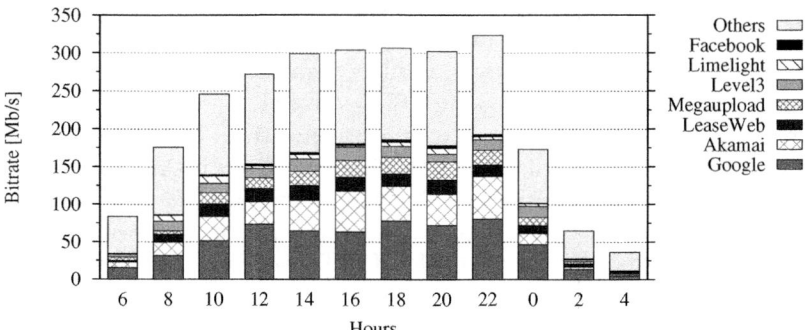

Fig. 1. HTTP volume breakdown for VP2 with respect to the organizations

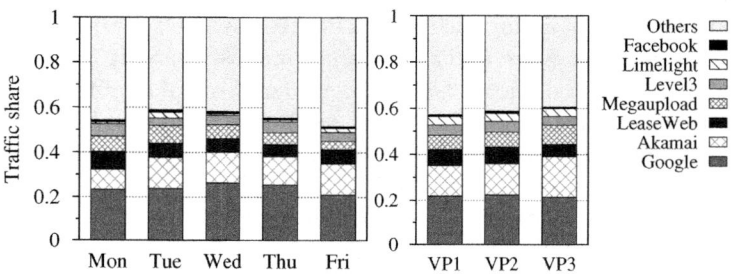

Fig. 2. Comparison of the HTTP traffic share during peak hours considering different days (left) and data sets (right)

observed. This is also true when considering different vantage points as reported in Fig. 2 (right).

Overall, we therefore notice that there is a limited impact on this kind of analysis when changing time, day, or vantage point. This has interesting implications on the possibility of monitoring the traffic from a few vantage points instead of deploying a capillary monitoring system. However, we expect the picture to be very different when changing ISP or nation, given we expect users' habits and cultural interest play an important role.

3.3   Content Characterization

To investigate and compare the characteristics of the content served by each organization we rely on the DPI capabilities of Tstat. In particular, we define five coarse classes of content: HTTPS/SSL, Facebook, File Hosting, Video Content and Generic web content.

Fig. 3 reports the breakdown of the volume of each organization with respect to the five classes, sorting organization in increasing fraction of Generic web. Most of the organizations serve Video Content. More precisely, Akamai hosts Vimeo, Megaupload and Leaseweb serve Megavideo, while Level3 and Limelight

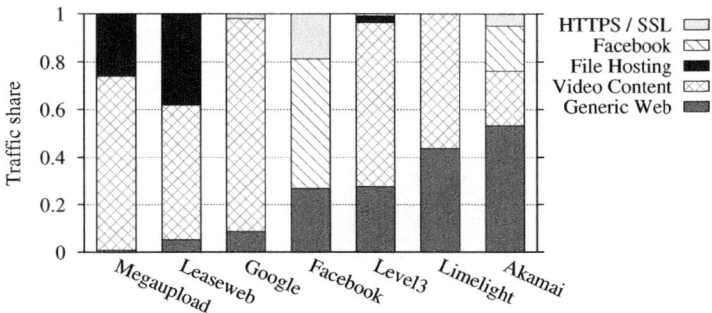

Fig. 3. Breakdown of the HTTP volume downloaded from each organization with respect to the type of content. Results refer to VP2 data set.

Table 3. Fraction of HTTPS/SSL traffic for each organization

Organization	%Bytes Jun	%Bytes Oct	%Flows Jun	%Flows Oct
Google	1.75	2.67	11.1	17.34
Akamai	3.46	2.95	10.55	17.6
Leaseweb	0.46	0.26	0.55	1.85
Level3	1.22	0.34	1.49	2.48
Limelight	0.05	0.78	1.82	2.16
Megaupload	-	-	-	-
Facebook	13.3	26.8	15.35	24.3

host several adult video sites, and Google hosts all YouTube videos. Considering File Hosting services, Level3 serves RapidShare while Megaupload and Leaseweb serve the files for the Megaupload site.

Interestingly, a small fraction of traffic is encrypted. To give more insight, Tab. 3 reports the fraction of volume and flows related to HTTPS/SSL traffic for each company. The results refer to VP2, for which a more recent data set has been collected during October 2011. Results show that encrypted traffic is significant only for Google, Facebook and Akamai. Both Google and Facebook are known to provide services for which the end user is allowed to opt for HTTPS as the default application protocol. In case a user has opted to encrypt its communications with Facebook, then all Facebook content will be provided over HTTPS, including content coming from the Akamai servers. This explains the relative large fraction of HTTPS/SSL flows served by Akamai.

Comparing the June and October data sets, we notice that encrypted traffic is increasing its share. For Facebook it doubled, with 26% of bytes and 24% of flows encrypted in October. This suggests that Facebook users are increasingly opting to enable the privacy/security functionality offered by the Social Network. The server cost to support encrypted traffic is thus increasing as well.

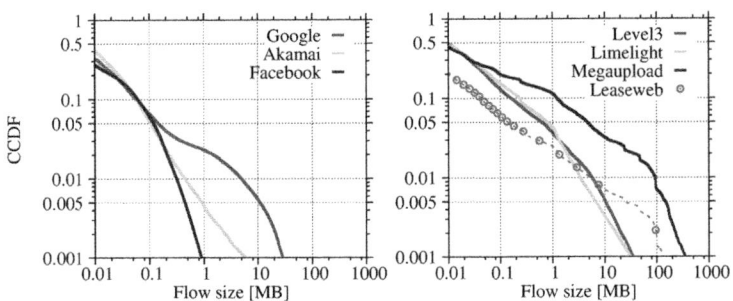

Fig. 4. CCDF of the server-to-client flow size for VP2

Finally, we focus our attention on the length of content that each organization has to handle, which reflects also the storage capacities they support. Fig. 4 reports the Complementary Cumulative Distribution Function (CCDF) of the amount of bytes the server sent to the client for the VP2 data set. As usual, results are identical for the other data sets too. As you can see, most of the flows are mice, i.e., they carry less than 10 kB. The fraction of mice flows ranges from 80% (Leaseweb) to 50% (Level3). Facebook, which hosts only small dynamic objects, has a negligible fraction of elephants, i.e., flows that carry more than 1 MB. For other organizations, more than 1% of the flows carry more than 1 MB. For Google, the larger number of elephants is due to the YouTube video flows. Megaupload is the organization that carries the largest number of elephants, since most of the content downloaded from its server is either due to complete movies or large archives. Surprisingly, 50% of the content served by Megaupload is less than 10 kB long, much smaller than average chunk size used by download managers. Overall, the flow length follows an heavy tailed distribution (note the log-log scale).

Considering the flow duration (not reported here due to the lack of space), for all the organizations but Megaupload and Facebook, the flows are shorter than 5 minutes, with more than 83% of the flows shorter than 1 minute, while Facebook and Megaupload present an heavy tail with more than 2% of the flows longer than 30 minutes. For Megaupload this reflects the length of the content downloaded from the server, while for Facebook this is due to an intentional use of persistent HTTP connections that are used to carry some services like chat or Ajax applications that update the web page dynamically.

4   How Organizations Handle Traffic

We now investigate on how large is the number of servers that have been contacted in our data set for each organization, and how traffic is balanced among servers. To augment the visibility on each organization servers, in this section all three data sets are aggregated and analyzed.

Table 4. Number of IPs and /24 subnets for each organization. Results refer to the aggregation of all data sets.

Organization	IP addresses		/24 Subnets	
	No.	(%)	No.	Top5 %Bytes
Google	3678	(0.76)	135	94.1
Akamai	10445	(2.16)	1255	86.1
Leaseweb	3833	(0.79)	546	80.0
Level3	1868	(0.39)	572	65.8
Limelight	1179	(0.24)	115	97.2
Megaupload	808	(0.17)	15	64.0
Facebook	338	(0.07)	27	74.5
Total	33798	(7.00)	4596	-

Tab. 4 reports the number of unique server IP addresses belonging to each organization that have been contacted by any customer. For each organization, the second column shows the total number of server IP addresses (referred as servers in the following for simplicity), the third column details their fraction over the total number of HTTP servers, and the forth column reports the number of /24 subnets the IP addresses belong to[3]. Considering each organization individually, we have also computed the traffic volume generated by each subnet. Ranking the subnets according to this measure, the last column shows the percentage of volume delivered by the top 5 subnets of each organization.

First of all, the total number of servers belonging to the selected organizations accounts to 7% of all HTTP servers. Yet, those 7% of servers are responsible for more than 60% of the HTTP traffic volume (see Tab. 2). Considering both the number of IP addresses and /24 subnets and how they are used, we can observe that the organizations pursue different approaches. Akamai is the organization with the highest number of servers and subnets, respectively 10445 and 1255. Google, which serves a similar amount of traffic as Akamai, uses one third of the servers, which are aggregated into one tenth of /24 subnets. This reflects the different architecture Akamai and Google have: small data centers spread over many ISPs [4] for Akamai, few large data centers for Google [5]. Limelight, another large CDN organization, follows a more concentrated deployment as well.

Facebook results the organization with the smallest number of IPs. Being most of Facebook static content actually served by Akamai, Facebook can handle the dynamic content of his social network by managing 338 servers only. Megaupload, another very specialized organization targeting file hosting services, shows a small number of servers and subnets as well.

Considering the last column of Tab. 4, we see that most of the traffic is actually served using only few subnets, with the five most used subnets accounting for more than 64% of the volume of each organization. Interestingly, the top 5

---

[3] The aggregation of servers in /24 subnets may not coincide with actual subnets. We consider it as an instrumental aggregation.

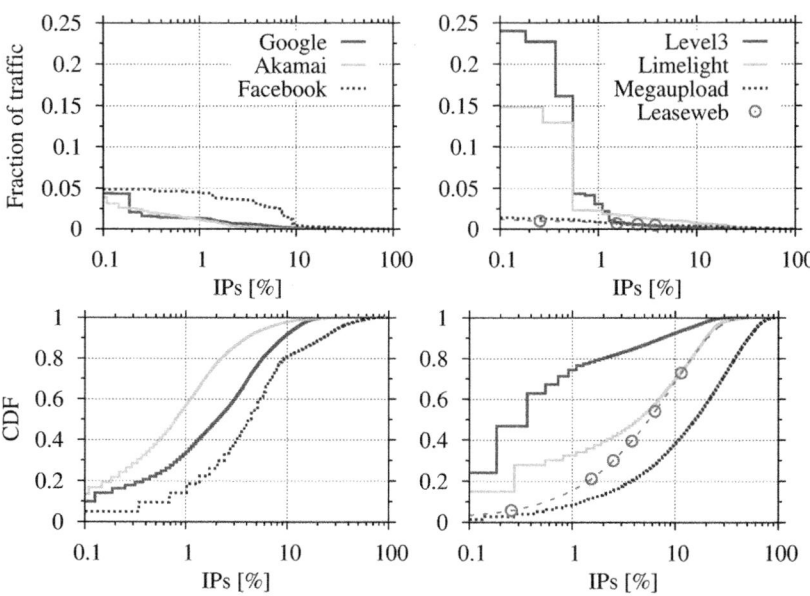

Fig. 5. Fraction of volume served by each IP (top) and the associated CDF (bottom) for each organization in VP2 data set

subnets of Limelight serves 97% despite we have found 115 subnets hosting its servers and being contacted by the customers.

To give more insight about how the traffic is concentrated, Fig. 5 shows the fraction of traffic generated by each single server. Servers have been ordered in decreasing fraction of traffic, and their rank has been normalized to the total number of servers in each organization. Top plots report the fraction of volume served for each IP, while bottom plots report the CDF. Several differences arise. Akamai tends to concentrate traffic on the same subset of server more than Google. Consider for example the fraction of servers that account for 80% of traffic. Akamai handle this with 2% of its servers. Google uses 5% of servers, while Facebook uses 10% of servers. In all cases, the top server handles about 5% of total volume. Considering other organizations, Level3 and Limelight follow a even more biased distribution of traffic, with the top three servers accounting for 63% and 31% of total traffic volume respectively. Finally, Megaupload and Leaseweb spread the traffic more uniformly among their servers.

The different concentration of traffic among the servers can be related to different causes: i) we observed that more than 3% of the users use third-party DNS resolvers as OpenDNS (1%) or Google's DNS (0.5%) which do not manage to redirect the users to the preferred data center, contrary to the ISP DNS servers [11]; ii) load balancing schemes are used to cope with day/night traffic shift redirecting the users to different servers according to the time of the day (more on this later on); iii) the content requested might be not available at the server contacted leading to application-layer redirections (e.g., YouTube [5]).

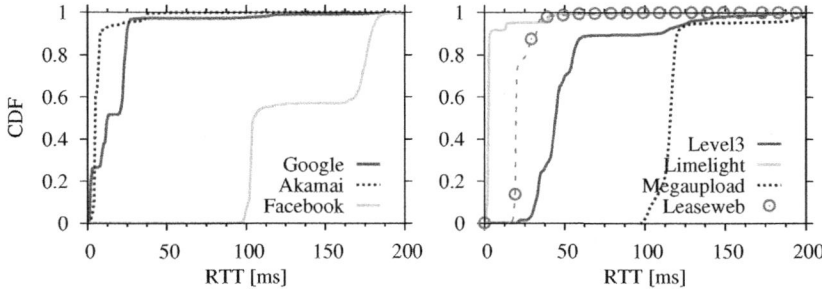

Fig. 6. Distribution of the minimum external RTT in VP2

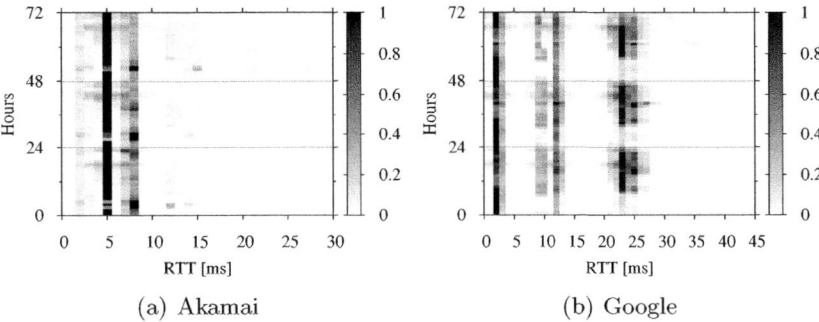

(a) Akamai   (b) Google

Fig. 7. Variation of the minimum external RTT over three days for VP2 data set

5   Performance Analysis

In this section we investigate the performance offered by the considered organizations. We start analyzing the Round Trip Time (RTT) as a measure of the server distance from the users hosts. We then focus on the bulk download rate for elephants.

5.1   Round Trip Time

The RTT is a typical measure that impacts both the TCP performance and the network cost. The smaller the RTT, the faster the TCP three-way handshake, the larger the download rate, and the less the network must handle the packet. Following the same methodology as in [5], we consider the *minimum external RTT* observed for each TCP flow over all valid RTT samples during the flow lifetime[4]. Only flows with at least 5 valid RTT samples are considered.

---

[4] The external RTT is measured considering the time elapsed between the data packet sent by a host inside the vantage point and the corresponding acknowledgement sent by the server.

Fig. 6 reports the CDF of the minimum external RTT observed from VP2 considering the whole week long data set. As expected the distributions present sharp knees corresponding to different servers positioned at different data centers. It follows immediately that different organizations are located at different distances from the vantage point. Flows going to Akamai, Limelight and about one third of Google servers are within few milliseconds far from the vantage point, i.e., very close to the ISP peering points; Level3 and the majority of Leaseweb flows are handled by servers in [20,50] ms range, i.e., possibly in some European country; at last, Megaupload and Facebook servers are found above 100ms, i.e., outside Europe. The long tail reflects the chance to contact one server which is not among the preferred location (recall Tab. 4).

More interestingly, some organizations leverage more than one server set (or data center) to handle the traffic. This is clearly the case for Facebook, for which two data centers are highlighted, the first at about 100ms, the second at about 160ms. Manually investigating their position, we uncover that the first data center is located in the New York area, while the second is located in the Bay Area. Interestingly, each of them serves about 50% of the traffic. Limelight and Level3 also show more than one (preferred) location. While Level3 balance the traffic among the different data locations, Limelight shows a higher preference to serve the requests from the closest data center. Finally, most of Akamai traffic is served by two data centers very close each other and to the ISP customers; recall indeed that 2% of Akamai servers manages 80% of traffic, see Fig. 5.

To give more insights about the traffic fluctuations during the day, we consider each hour long data set, computing the histogram of minimum external RTT samples in bins of 1 ms. Each histogram is then normalized with respect to its largest value. The heat maps reported in Fig. 7 shows the density of each bin for each histogram considering three days. The darker the color, the higher the fraction of flows that falls in that RTT bin. Consider first the left plot which refers to Akamai. We notice little variations over time, with most of the flows experiencing either 5 or 7 ms RTT, i.e., corresponding to the two data centers. Other organizations but Google show similar results with no dependency over time of the RTT.

Consider instead the right plot which refers to Google. It presents three possible values centered at about 3, 12, 23 ms, each corresponding to a different data center. Those correspond to the knees in Fig. 6. Interestingly, we notice that at different hours a different fraction of traffic is sent to different data center. This unveils the presence of routing policies that direct the traffic based not only on the distance but also on the time of day, i.e., the current system load. In fact, during the night when the number of requests to be served is the lowest, the "preferred" servers are the closest ones. This is not true during the day. A similar effect has been found in [5] when dealing with YouTube traffic. According to our findings, this dynamic routing affects all Google CDNs and not only YouTube traffic.

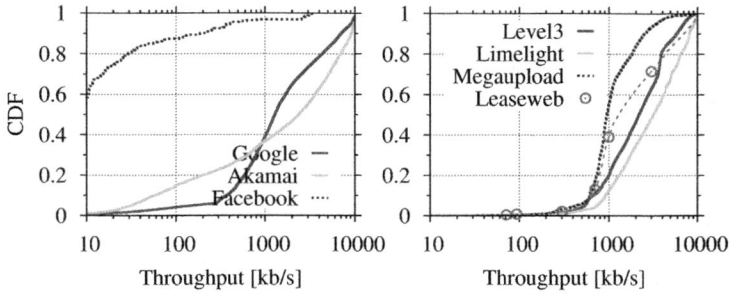

Fig. 8. Download rate of the flows with more than 1 MB for VP2

5.2   Download Throughput

We now focus on the bulk download rate defined as the total amount of bytes received from the server over the time elapsed between the first and last data segment. Only flows that carry more than 1 MB are considered. Fig. 8 shows the CDF aggregating the whole week in VP2 data set.

All the organization, except for Facebook and Akamai, present a similar distribution with a knee around 300-500 kb/s and a heavy tail. Considering the 60th percentile, Limelight offers the best performance, followed by Level3, Google and Leaseweb with 3 Mb/s, 2 Mb/s and 1 Mb/s respectively. Megaupload presents a shorter tail with a higher fraction of flows between 700 kb/s and 1 Mb/s. This is due to the throttling imposed by Megaupload on download rate to push customers to buy a premium access [6]. Finally, Facebook flows are much slower than others, with only 5% of the flows having more than 1 Mb/s. This is because Facebook is serving small objects using persistent HTTP connections for which the bulk download rate is a meaningless index. Most of Facebook flows are indeed short lived.

Interestingly, Akamai throughput follows a different distribution with 40% of flows that cannot exceed 1 Mb/s, and 40% exceeding 4 Mb/s instead. Investigating further, we have seen that this might be due to the probability of finding the content on the contacted server. If the server has already the content available, then the download is very fast. Instead, if the server has to retrieve the content, the download rate is much more limited. This suggests some form of congestion on the cache back-end rather than on the path from the server to the end customer. We have investigated if there is any time correlation, e.g., if the throughput increases during off-peak time, but we did not observe any correlation.

In particular, for organizations specialised in video distribution, i.e., Google, Akamai, Level3, the values of throughput can be biased by videos delivered with progressive download techniques that results in a bandwidth throttling by the video server.

6   Conclusions

In this paper we presented a first set of measurements that focus on the observation of HTTP traffic generated by the top player organization in the current Internet. We consider a one-week long data set collected from three different vantage points inside the same operator, analyzing the traffic from the customers' point of view.

After identifying the top players, we dug into the characteristics of each organization. We have unveiled how large is the number of servers each organization manages, how traffic is routed to which server or data center, how far those servers are from the customers, and which performance they offered. Results we collected are an interesting starting point to understand how today Internet traffic is delivered to customers.

References

1. Labovitz, C., Iekel-Johnson, S., McPherson, D., Oberheide, J., Jahanian, F.: Internet Inter-domain Traffic. In: Proceedings of the ACM SIGCOMM 2010 Conference, pp. 75–86. ACM, New York (2010)
2. Su, A.-J., Choffnes, D.R., Kuzmanovic, A., Bustamante, F.E.: Drafting Behind Akamai (Travelocity-based Detouring). SIGCOMM Comput. Commun. Rev. 36, 435–446 (2006)
3. Triukose, S., Wen, Z., Rabinovich, M.: Measuring a Commercial Content Delivery Network. In: Proceedings of the 20th International Conference on World Wide Web, WWW 2011, pp. 467–476. ACM, New York (2011)
4. Nygren, E., Sitaraman, R.K., Sun, J.: The Akamai Network: a Platform for High-performance Internet Applications. SIGOPS Oper. Syst. Rev. 44, 2–19 (2010)
5. Torres, R., Finamore, A., Kim, J., Mellia, M., Munafò, M., Rao, S.: Dissecting Video Server Selection Strategies in the YouTube CDN. In: IEEE ICDCS (2011)
6. Mahanti, A., Williamson, C., Carlsson, N., Arlitt, M., Mahanti, A.: Characterizing the File Hosting Ecosystem: A View from the Edge. Perform. Eval. 68, 1085–1102 (2011)
7. Finamore, A., Mellia, M., Meo, M., Munafò, M.M., Rossi, D.: Experiences of Internet Traffic Monitoring with Tstat. IEEE Network 25(3), 8–14 (2011)
8. Bonfiglio, D., Mellia, M., Meo, M., Rossi, D., Tofanelli, P.: Revealing Skype Traffic: When Randomness Plays with You. SIGCOMM Comput. Commun. Rev. 37, 37–48 (2007)
9. Finamore, A., Mellia, M., Meo, M., Rossi, D.: Kiss: Stochastic Packet Inspection Classifier for UDP Traffic. IEEE/ACM Transactions on Networking 18(5), 1505–1515 (2010)
10. Maxmind Geoip Tool, http://www.maxmind.com/app/organization
11. Ager, B., Mühlbauer, W., Smaragdakis, G., Uhlig, S.: Comparing DNS Resolvers in the Wild. In: ACM IMC (2010)

Internet Access Traffic Measurement and Analysis

Steffen Gebert[1], Rastin Pries[1], Daniel Schlosser[1], and Klaus Heck[2]

[1] University of Würzburg, Institute of Computer Science, Germany
`{steffen.gebert,pries,schlosser}@informatik.uni-wuerzburg.de`
[2] Hotzone GmbH
`heck@hotzone.de`

Abstract. The fast changing application types and their behavior require consecutive measurements of access networks. In this paper, we present the results of a 14-day measurement in an access network connecting 600 users with the Internet. Our application classification reveals a trend back to HTTP traffic, underlines the immense usage of flash videos, and unveils a participant of a Botnet. In addition, flow and user statistics are presented, which resulting traffic models can be used for simulation and emulation of access networks.

Keywords: Traffic measurement, traffic classification, flow statistics, broadband wireless access network.

1 Introduction and Background

As the web evolves, Internet usage patterns change heavily. For the simulation and development of new protocols and routing algorithms, it is important to have accurate and up-to-date Internet traffic models that describe the user behavior and network usage. For network administrators, it is crucial to have knowledge of what is happening in their network. Additionally, consecutive measurements help to identify bottlenecks, network misconfiguration, and also misuse.

A few years ago, Peer-to-Peer (P2P) file sharing was with about 40% the dominating service in the Internet [1]. Since the movie industry is pursuing users of file sharing software distributing copyright protected content, file download sites like RapidShare started attracting more people. Besides file download sites, YouTube's popularity accounts for an additional enormous amount of traffic that is transferred using HTTP. These observations were already made by Wamser et al. [1] back in 2008.

In this paper, we evaluate the same access network as in [1]. However, in the meantime, the observed network has grown and more than doubled its size, connecting about 600 users with the Internet. The increased size of the network required a new design of our measurement architecture to be able to collect and process the increased amount of data in real-time. We evaluate whether the high share of P2P still applies or if any other trend can be observed. In contrast to the measurements published in [1], we also evaluate flow statistics and present a user statistic.

The rest of this paper is organized as follows. Section 2 presents the work related to this paper. In Section 3, we describe the observed network and the measurement scenario. Section 4 shows the results of the 14 days lasting measurement. Finally, conclusions are drawn in Section 5.

A. Pescapè, L. Salgarelli, and X. Dimitropoulos (Eds.): TMA 2012, LNCS 7189, pp. 29–42, 2012.

2 Related Work

An analysis of 20,000 residential Digital Subscriber Line (DSL) customers is described by Maier et al. [2], which took place in 2008/09. Analyzing DSL session information, available through the ISP's RADIUS authentication server, revealed an average duration of 20-30 min per online-session. Observing the transfer times of IP packets exposes that most of the delay, in median 46 ms, is created between the user's device and the DSL Access Multiplexer (DSLAM), compared to 17 ms to transfer it to the connection end point (mostly a server accessed through the core network). HTTP dominated the traffic volume with a share of almost 60%, compared to 14% of P2P traffic. Furthermore, a distribution of the different MIME types of HTTP transfers is presented.

Classifying flows by only looking at connection patterns between hosts (host behavior-based classification) is done by BLINC [3]. By setting up connection graphs, so called *graphlets*, and matching observed traffic against them, a categorization of single hosts into different classes is made.

Szabó et al. [4] applied multiple classification methods to minimize the amount of unknown traffic and to compare the results of the different methods. The outcome was that every classification type and engine has its strengths and weaknesses. The final decision takes the different methods into account. If no majority decision can be made, classification results from the methods with the highest precision (signature- or port-based) are preferred over results achieved with vague heuristics.

Finamore et al. [5, 6] present their experience of developing and using their traffic measurement software *Tstat* during the past 10 years. From the technical perspective, they highlight design goals of measurement software coping with Gigabit speed and present statistics, where hardware capture cards cause an immense release of CPU resources.[1] By combining several classification techniques, including different DPI variants and a behavioral classifier, classification rates of more than 80% of data value were reached. Outgoing from four vantage points at end-customer ISPs with up to 15,000 connected users located in Italy, Poland, and Hungary they demonstrate that application usage is subject to variation by the observed country. Presented statistics range from May 2009 to December of 2010, thus also cover our measurement time frame of June/July 2010. At this time, in their measurement, eMule file sharing accounts for 15-35% of total Italian traffic, its share in Poland was around 1-2%, and not even visible in the Hungarian statistics. In contrast, Bittorrent was very popular in Hungary with around 35% traffic share respectively 20% in Poland. The majority of Polish traffic was observed to be caused by HTTP-based traffic, proportioned into each 15% of video streaming and downloads of file hosting services and 40% of the total volume general HTTP traffic. In contrast, its share in Italian and Hungarian is only half as high with 30-40%. At this particular time in June 2010 they observe - contrary to previous trends - a turn in P2P traffic share, which started increasing again.

Our study fits very well to the work of Finamore et al. [5], which presents data of other European countries, while we now provide numbers for German Internet usage. In addition to general application statistics, we focus on flow characteristics in the second part of this paper.

---

[1] According to the figures from 100% load to less than 10% at a link utilization of 300 Mbps.

3 Measurement Scenario and Methodology

In this section, we describe the measurement scenario as well as our measurement framework with its software components, which used during this evaluation.

3.1 Setup

In cooperation with a German broadband wireless access provider, we observed an access network to which about 600 households are connected. The network connects two housing complexes with the Internet through another carrier ISP. Complex 1 provides wireless access for about 250 households through numerous Cisco Access Points. Complex 2 provides wired access for about 350 households using the in-house LAN installation. Complex 2 is connected to Complex 1 using an 802.11n WiFi link.

The connection to the Internet carrier is established using another 802.11n WiFi link, which is limited to 30 Mbps per direction. Although this seems to be a quite slow connection, it is according to the provider completely sufficient to support all 600 households. All three links are connected in an exchange network, located in Complex 1. Figure 1 illustrates the established links and the position of our measurement point in the exchange network, right before the WiFi link to the carrier.

Besides artificial limitation of the WiFi link to the carrier, a class-based traffic shaping from Cisco [7] is applied. Traffic shaping can limit the bandwidth available for different services. Contrary to dropping packets, shaping limits the bit rate more softly. Through the implementation of a token bucket filter, packets are sent only at a specified maximum rate. Downside is an increased delay, as packets have to wait for some time, in times of non empty queues. When the incoming rate is below the shaping limit, no delay is added, as packets are forwarded immediately. Queued shaping can only limit

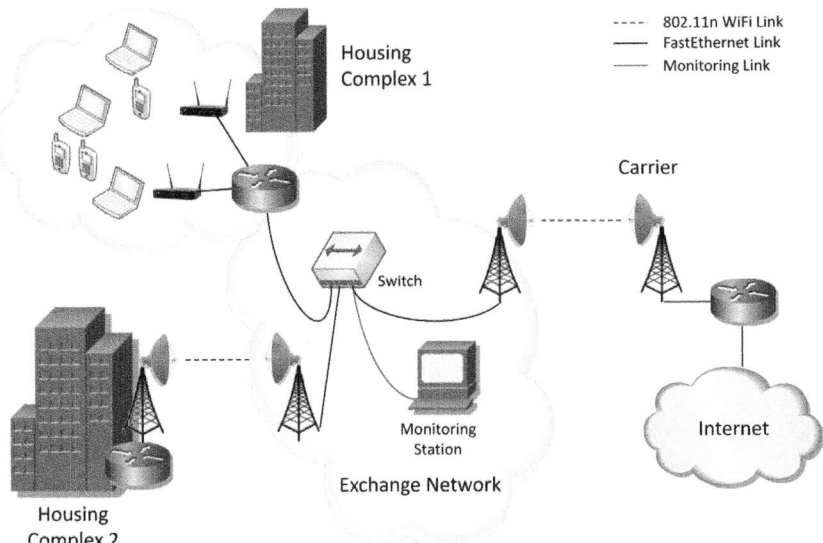

Fig. 1. Network and measurement setup

the average rate, not the peak rate. If the queue is empty, the rate can be way higher for a short time frame, until the buffer is filled up.

The ISP's shaping configuration is as follows. Based on a white-list, traffic is added to a *high-priority class*. This class is allowed to consume up to 22 Mbps per direction and includes the most frequently used and "known-good" protocols like HTTP, HTTPS, email, SSH, Skype, or video streaming applications. They are identified using Cisco's content filtering engine. Furthermore, on user request, ports are added to the high-priority list, to cope with limitations of the classification engine. All other protocols that are not identified to be high-priority, are put into a *default class*. This especially matches P2P file sharing software like Bittorrent. The default class is limited to an average bandwidth of 1 Mbps per direction. This way, excessive use of file sharing and negative influence on real-time services is prevented according to the provider's opinion. However, during our measurements, we observed that the limitation with Cisco's content filtering engine does not work properly.

Clients get public IP addresses with a short DHCP lease timeout of 30 minutes assigned. Therefore, IPs are no valid identifier to track a particular user, as a reassignment of the address after an inactivity of at least 30 minutes is very likely. As the ISP firewall allows incoming connections, people are directly exposed to the Internet and able to run publicly reachable services.

3.2   Measurement Software

Our measurements are done using our custom measurement tool PALM, which is targeted to run on commodity hardware. PALM stands for *PAcket Level Measurements* and is designed to generate packet and flow level statistics. Although packet level statistics are available, we focus in this publication on flow level statistics. We designed PALM in such a way that a throughput rate of at least 50 Mbps symmetric with an average packet rate of more than 50k packets per second can be measured, classified, and stored.

We could have used community-oriented measurement software such as TIE [8], but decided to write our own software because we were able to reuse parts of our previous software applied in [1].

Figure 2 depicts the architecture of PALM. Its three major components *Collector*, *Processor*, and *Store* are separated into different threads for use of multiple CPU cores. All components are highly decoupled to stay flexible for exchanging single implementations and for further expansion, e.g. through additional classification engines.

Packet processing starts with a *Collector*, which reads packets either from the network interface or a PCAP file. We make use of the *SharpPcap* library, which provides

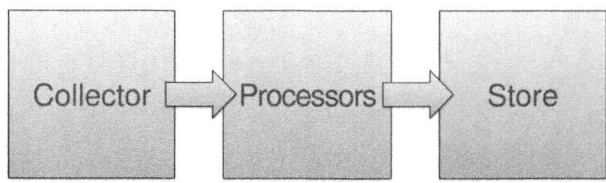

Fig. 2. Measurement software architecture

a C# interface to the *libpcap* library. In the next step, several *Processors* are executed on a per packet basis, from which we elaborate only on the most important ones.

OpenDPI: This processor makes use of OpenDPI, an open-source Deep Packet Inspection (DPI) library, for classification of packets. Instead of enforcing a classification rate of near to 100% bringing the risk of an immense number of false classifications, we prefer a more defensive strategy and relinquish vague heuristics. Table 1 lists the most important applications and their arrangement in different categories.

Table 1. Application categories of the packet classification

Category	Protocol	Category	Protocol
HTTP	HTTP	MULTIMEDIA	Flash
	HTTPS		RTP
	DirectDownloadLink		SHOUTcast
IM	Jabber		PPStream
	Yahoo		MPEG
	ICQ		Windowsmedia
	MSN	OTHER	IPSec
	IRC		DHCP
MAIL	IMAP		SNMP
	POP3		NETBIOS
	SMTP		NTP
P2P	Bittorrent		SSH
	Gnutella		DNS
	PANDO		ICMP
		UNKNOWN	*unknown*

IpAnonymizer: Very important while dealing with user data is anonymization. We run a basic *IpAnonymizer*, which scrambles IP addresses. In case of higher demands, e.g. prefix-preserving anonymization, we think of extending PALM by making use of external libraries like *Crypto-PAn* or *PktAnon* [9].

FlowManager: Through this processor, packets having the same quintuple (source/ destination IP/port and protocol) are aggregated into flows. We treat a flow as unidirectional data transfer. Thus, in most cases an opponent flow with exchanged source/destination IP/port exists. As suggested by CAIDA, we use an inactive-timeout of 64 seconds for flow termination [10]. By executing flow aging every 60 seconds in a background thread, longer inactivity of a flow (up to 124 seconds) could prevent termination of the first and starting of a second one with the same quintuple and merge them to one flow. However, we do not count this as a disadvantage for our measurements.

After real-time analysis and anonymization, packet and flow data is saved by a *Store*. We made good experience with our *MySQL* store, which is able to save data at a very high rate of more than 100k records (packets or flows) per second by using bulk inserts[2]. The

---

[2] Bulk inserts can be done by using the *LOAD DATA INFILE* command.

advantage of using a Relational Database Management System (RDBMS) like MySQL is that we are able to directly gain statistics by executing SQL queries, also via software like Matlab[3], without any intermediate conversion steps.

3.3   Measurement Duration

Our measurements took place in June and July of 2010. For the evaluation presented in this work, we use data of 14 days. Table 2 lists statistics about the collected data.

Table 2. Measurement statistics

Beginning of measurement	2010-06-28
End of measurement	2010-07-12
Number of packets	4,946,071,829
Number of flows	202,429,622
Amount of data observed	3.297 TB
Number of households (acc. to ISP)	ca. 600
Carrier link speed	30 Mbps

4   Measurement Results

In the following, we present our measurement results retrieved from a 14 days measurement period. During these two weeks, 3.3 TByte were observed. This section is divided into daytime statistics, application usage, flow statistics, and usage statistics.

4.1   Daytime Statistics

We start with investigating the variation of Internet usage during the day. Observed traffic patterns dependent on the ISP's customers type, such as home or business users.

The network, which we are observing, mostly supplies smaller households. Figure 3(a) shows the daytime utilization of the network link in a 5-minute interval averaged over the 14 days of measurement. The minimum usage is observed around 5:00 to 6:00 in the morning, when most people are sleeping. The throughput increases steadily over the whole day and reaches its maximum at about 22:00. Such observations of a three- to fourfold throughput during late hours compared to early mornings are perfectly in line with the values reported for home users in [6].

To get an insight from which applications the traffic is originating, Figure 3(b) presents the daily traffic fluctuation, grouped by application type, omitting less important categories. Peer-to-Peer applications cause an almost constant bit rate of 3-4 Mbps. This reveals two facts: On the one hand, the ISP's traffic shaping is not working that perfect. This can be caused either by P2P applications using ports prioritized by the traffic shaping and thus circumventing the 1 Mbps per direction limit, or the configured "average" limit of Cisco's shaping functionality is really very inaccurate. The second

---

[3] Using the *mYm* extension for Matlab.

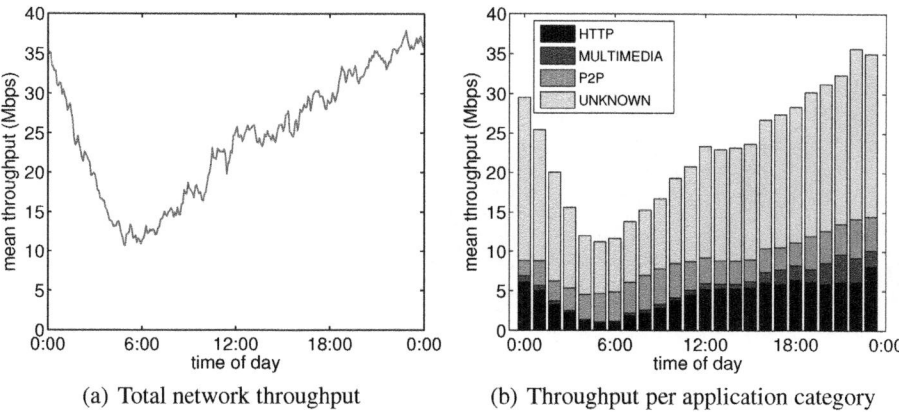

(a) Total network throughput   (b) Throughput per application category

Fig. 3. Mean daily traffic fluctuations

fact shown by this statistics is that users of file sharing software often do not switch off their computers during night.

Contrary to P2P traffic, HTTP is heavily influenced by the day-night fluctuation. Except for large file downloads, HTTP is interactive traffic, caused by the user accessing web pages. In line with the expectations, the minimum usage is observed between 5:00 and 6:00, when only a few people are actively using their PCs. HTTP throughput increases steadily between 7:00 and 12:00 from below 1 Mbps to 5-6 Mbps. For the rest of the day, the throughput stays almost constant. An absolute maximum is reached in the hour between 23:00 and 0:00, where its average traffic is at 8.1 Mbps. The factor between the throughput at 6:00 in the morning and during the day at 12:00 and 18:00 is 1:5.4 respectively 1:6.5. The characteristic that HTTP traffic is mostly interactive traffic plays a major role in the day-night fluctuation of the overall network usage.

Streaming and video download traffic is even more subject to variation. Again, the minimum usage is observed at 5:00. The utilization increases, until its maximum is reached at 20:00. Especially in the evening, more intense use of streaming is observed. The throughput relations between 0.18 Mbps at 6:00 compared to 0.7 Mbps at 12:00 and 1.78 Mbps at 18:00 are even more diverse than of HTTP traffic. As pointed out, HTTP stagnates about noon. Increased usage of multimedia applications is mainly responsible for the further increase of the network throughput during the rest of the day.

4.2 Application Usage

We continue with presenting the top-15 protocols regarding the amount of transferred data and number of flows in Figure 4. The y-axis show a logarithmic scale.

We observe HTTP traffic as the protocol that is used for most data transfers with a share of 18.4%. OpenDPI can return more fine-grained results for some protocols using HTTP as transport, like Flash or downloads of Windowsmedia (wmv) and MPEG video files. It also comes with detection rules for the most popular DirectDownload hosters, like Rapidshare. Putting them altogether and counting all HTTP transfers increases the share to 22.6%, which makes HTTP the dominating protocol.

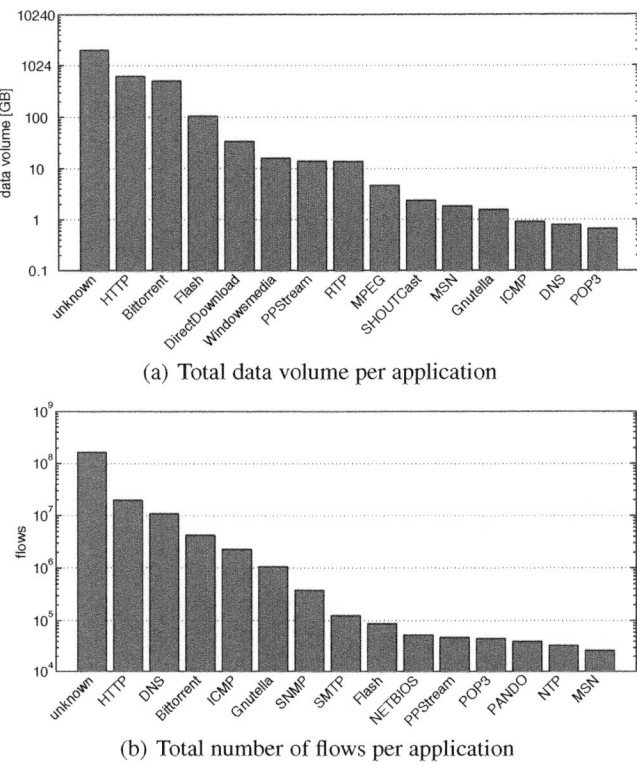

(a) Total data volume per application

(b) Total number of flows per application

Fig. 4. Total data volume and number of flows per application

While [6] reports 12-15% traffic share of video streaming for Italian and Hungarian customers and 20% for Polish users, we observe with 3.0% a lot less. The share of Flash on total HTTP traffic is for us with 17% also smaller compared to their observations. Furthermore, [6] reports an increase of 10% of Flash traffic starting from January to July of 2010, which since then stagnates. Very likely, this trend is shifted by a few months for the German users that which we observed.

The second most seen protocol is the P2P file sharing protocol Bittorrent. As explained in Section 3.1, Bittorrent traffic is throttled in the observed network. Although P2P file sharing usage decreased in favor of DirectDownload hosters over the last years [1], the share of 14.8% is immense. On average, more than 35 GB per day are Bittorrent data transfers. Compared to values for other European countries [6], our German users show similar usage characteristics as users from Poland. However, as mentioned before, traffic is throttled in the observed network and also very likely a portion of the unidentified traffic is caused by P2P software, which we well discuss in following.

The observed amount of flows classified as *unknown* is tremendous. OpenDPI was not able to classify 80.1% of the flows, responsible for 64% of the traffic volume. More than 46% of those flows have no inverse flow, which gives a few options of guessing their origin, and will be discussed in following. First, if the connection uses TCP, these flows must have been unsuccessful connection attempts. As the first SYN packet contains no payload, it can not be identified by our signature-based classification. In total, 8.5% of the unsuccessful classified flows without an inverse flow are failed TCP connection attempts. UDP uses no connection establishment and the first packet already contains payload data. It is possible that the one-way UDP connection was arranged using another control connection or was defined by configuration. *SNMP Traps* or other monitoring data are examples, where no feedback is expected and data is sent without receiving any response packets. However, this is very unlikely in the Internet and only possible for sending data to servers, in case of clients behind NAT gateways.

Thus, we assume that most one-way conversations are failing connection attempts. These are most likely P2P applications, which try to find other peers, but fail due to the peer having already disconnected, or the packet being blocked by the target host's firewall. We measure that 91.6% of those one-way UDP flows consist of only one single packet, which proves the theory of being mostly connection attempts. Altogether 98.7% contain less than five packets and even 99.6% of the flows consist of less than ten packets. This leads to the conclusion that almost no serious one-way connections are used. We conclude that the main part of the mentioned 46% of all flows, which have been classified as unknown, is caused by failed connection attempts, which our Deep Packet Inspection is not able to identify. The remaining share of unsuccessfully classified flows are most likely caused by P2P applications and encrypted traffic.

Further exploration of *unknown* traffic based on port numbers reveals the following:

– Traffic originating from TCP port 80 (HTTP) causes the major share of 57.7% of the unknown traffic amount. As we found no obvious mistakes in OpenDPI's rule set, we assume that this is non-HTTP traffic tunneled through port 80. As pointed out in Section 3.1, the ISP's traffic shaping prioritizes traffic based on white-listed ports. We evaluated traffic tunneled through port 80 being classified as non-HTTP protocol and e.g. found 820 MB of Bittorrent and 1.3 GB of MSN messenger traffic.
– Port 13838, both TCP and UDP, is used by exactly one customer IP address for incoming connections during the two weeks. While the UDP flows account for 1.33% of unknown flows and 1.60% of unknown bytes, the TCP flows account for only 0.13% of unknown flows, but 6.13% of unknown traffic. On average, port 13838 TCP flows are 10 times bigger than the UDP flows. We have the feeling that this is a P2P application that uses UDP for signaling and TCP for data transfers.
– TCP Port 443 (HTTPS) accounts for 1.90% of unknown bytes and 0.69% of flows. Although encrypted, SSL connections can be identified during the handshake. OpenDPI ships with a signature, however this either does not support all SSL versions or, again, some applications use the prioritized port 443 for tunneling other traffic.
– Except these ports, we found only TCP ports 554 and 182 as port numbers below 10,000 in the 20 port numbers responsible for most of the unknown traffic. Port 182 is registered port for *Unisys Audit SITP*, 554 for *kshell* (Kerberos). Thus, we assume most of the unidentified traffic being P2P transfers using dynamic port numbers.

We stick to this as explanation for the high amount of unknown traffic and use the numbers gained by successful classifications from OpenDPI.

Besides OpenDPI, we also implemented and evaluated a *L7-filter* Deep-Packet-Inspection processor, which could help lower the amount of unknown traffic. However, we experienced an immense slowdown of the classification by a factor of 5-10. In order to reach our performance goal, we sticked to solely OpenDPI for this measurement.

Another thing, which attracted our attention, was the number of SMTP flows: One outgoing SMTP flow usually means sending one email. Observing 125,000 SMTP flows during the 14 days would mean that each of the 600 users writes more than seven mails per day. As many people prefer Webmail interfaces or use Facebook for their communication, this number seems way too high - given the fact that we the clients are mostly smaller households. Calculating the number of SMTP flows reveals that one IP address is responsible for 98,150 of these 125,000 flows, distributed over the the whole measurement time. This immense number of email sending attempts must be caused by a virus infection of this particular host. It is very likely that this host is part of a Botnet and acts as a sender of spam emails.

4.3   Flow Sizes per Application

In Figure 5, cumulative distribution functions (CDFs) are shown for the most frequently observed categories. OTHER combines all successfully classified flows together that do not fit into one of the other categories. DNS traffic, which is associated with this category, is responsible for 18.7% of all flows and mainly consists of one packet per flow. DNS is one of the reasons for this category containing more smaller flows, so called *mice*, than other categories. Almost 42% of the flows are less than 100 bytes in size and more than 81% are below 310 bytes.

The left part of the HTTP flow size distribution is caused by upstream flows, while the right part, containing the larger flows, is typically caused by downstream flows. In contrast to downstream flows containing the real payload data, upstream flows mostly contain only the HTTP request header and TCP acknowledgments (ACKs). The number of ACKs depend on the retrieved payload size. The size of the request header is influenced by the used browser, transferred cookie data, and some other information, including the length of the requested file name. Thus, the distribution starts very smooth, compared to the other application protocol categories, which show more discrete steps.

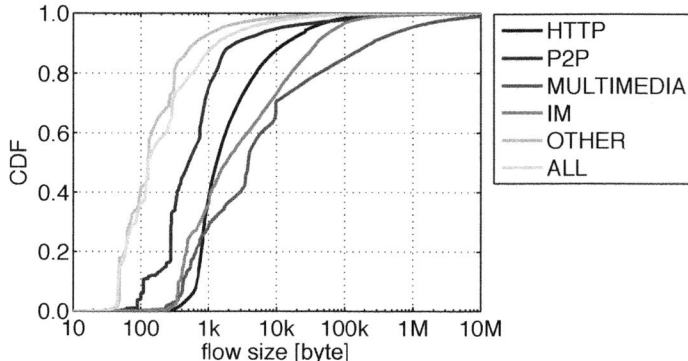

Fig. 5. Flow size distribution per application category

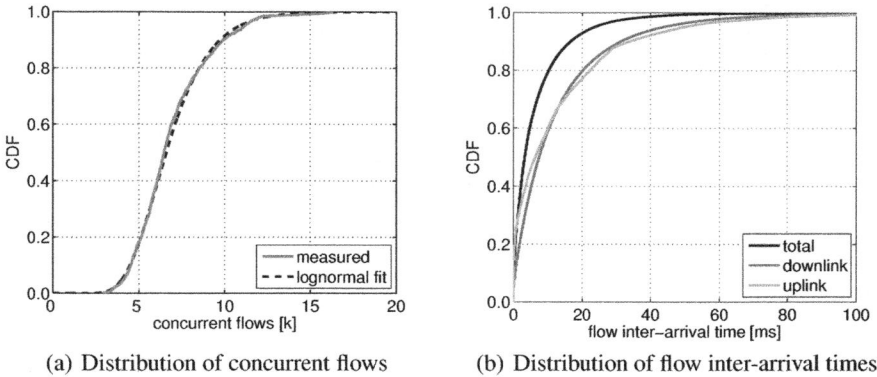

(a) Distribution of concurrent flows   (b) Distribution of flow inter-arrival times

Fig. 6. Distributions of concurrent flows and flow inter-arrival times

4.4 Concurrent Flows

The number of concurrent active flows has an influence on the required processing power of network devices operating on flow level. Besides of statistics collection, e.g. like on a NetFlow probe, the number of concurrent active flows gets an interesting measure with recent work on flow-based routing or switching, like OpenFlow [11].

For this evaluation, we calculate the number of concurrent active flows once every minute during one day. Figure 6(a) shows the resulting CDF. In 90% of the time, between 4,100 and 11,200 flows are simultaneously active. During 9.9% of the observed points of time, 10,000 and more concurrent flows are observed, in 0.9% even more than 15,000. The maximum number of concurrent flows is 16,290. The lognormal distribution with parameters $\mu = 8.79473$, $\sigma = 0.302074$ fits the measured distribution with a mean of 6907.45 concurrent flows best.

4.5 Flow Inter-Arrival Times

For the same day, we present the distribution of flow inter-arrival times (IAT) in Figure 6(b), both for the total traffic and split into uplink/downlink direction.

On average the next flow arrives within the next 3.23 ms. For egress traffic, a new flow is created every 6.3 ms and every 7.5 ms for ingress traffic. An inter-arrival time of less than 10 ms is observed for 60.0% of egress, and 59.0% of ingress traffic. For IATs of combined traffic directions, new flows arrive in less than 10 ms in 79.4% of all cases. In less than 10% of the cases, new flows arrive with delays higher than 34.0 ms for upstream, 31.5 ms for downstream, and 16.5 ms for combined traffic. These values are again useful as input for simulations of an access network or while dimensioning devices operating on a per-flow level, like the mentioned OpenFlow devices.

In order to figure out, which distribution the overall flow inter-arrival times are following, we fitted the CDFs with the exponential, gamma, and lognormal distributions. Table 3 shows them together with the optimal parameters. While the lognormal distribution does not reflect the observed IAT distribution very well, the exponential distribution and especially the gamma distribution show an accurate fit.

Table 3. Applied distribution functions for flow inter-arrival times (in ms)

Distribution	Parameters	Mean	Variance
Exponential	$\lambda = 6.5277$	6.5277	42.6107
Lognormal	$\mu = 0.5295,\ \sigma = 2.5058$	39.2175	818811
Gamma	$k = 0.4754,\ \theta = 13.7300$	6.5277	89.6254
Measurement		6.5277	87.0107

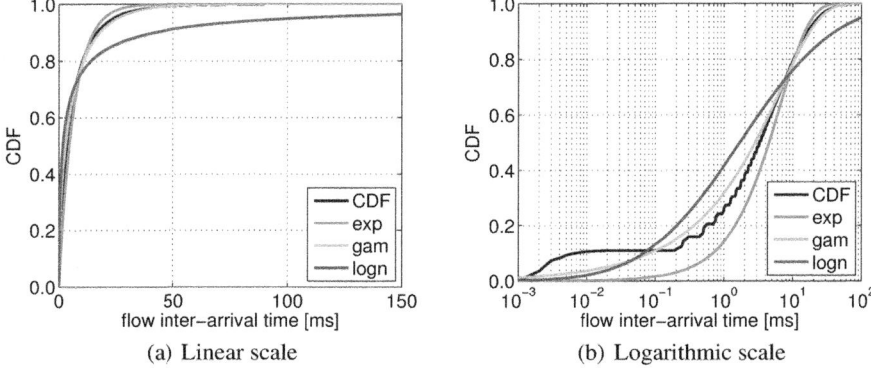

(a) Linear scale   (b) Logarithmic scale

Fig. 7. Fitting of flow inter-arrival times distribution

The exponential distribution function seems to fit pretty well according to Figure 7(a). However, looking at the differences using a logarithmic x-scale in Figure 7(b) reveals that especially in the smaller time distances, its failure is way larger than the mean error of the gamma distribution. As with a probability of 50%, the time distance between two new flows is below 3.3 ms, a larger failure of the fitting distribution for smaller IATs makes the disadvantage of the exponential distribution even more appealing. Using a gamma distribution to estimate flow arrivals is also preferred by [12].

4.6   Internet Usage Statistics

As the ISP has a very short DHCP lease times of only 30 minutes configured, it is hardly possible to track after switching their PCs off and on. To get an impression of the client's traffic consumption, we evaluate logging data of the ISP's billing system. Figure 8 shows the distribution of the amount of traffic caused by the users during one month, separated into upstream, downstream and total traffic.

After a steep increase until about 2 GB transfer volume, the curve flattens heavily. This is the transition from users, accessing the Internet mainly for web browsing, email, and very likely social networking, towards the user category, making intense use of video or audio streaming sites and services. Already 63.5% of the users cause a traffic amount of more than 10 GB per month, 17.5% more than 40 GB. Within this region is the fluent transition towards the group of power users, making excessive use of downloading, file sharing, video streaming, and other bandwidth consuming applications.

Table 4. Gamma distribution parameters for modeling of monthly user traffic consumption

Direction	Parameters	Mean
Downstream	$k = 0.622127\ \theta = 29.4298$	18.3091 GB
Upstream	$k = 0.391863\ \theta = 13.1583$	5.15626 GB
Combined	$k = 0.589397\ \theta = 39.8124$	23.4653 GB

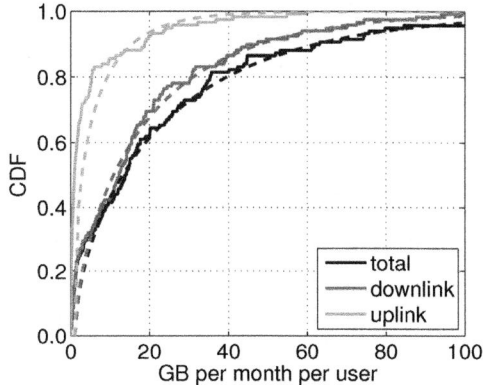

Fig. 8. Total traffic consumption per user

A monthly transfer volume of more than 100 GB is observed for 4% of the ISP's users. The largest observed volume was 135 GB.

An average traffic volume of 12 GB for German households, respectively 19 GB for US households, in 2009 was reported by [13]. The mostly single households, which we observed, make more intense use of their Internet connection and cause a higher traffic volume. We measured an average traffic volume of 21.9 GB per month per user. Table 4 lists the parameters of the gamma distribution, which best fits the user statistics.

5 Conclusion

In this paper, we presented the results of a 14 days Internet access network measurement. In contrast to our previous publication [1], we had to completely redesign the measurement infrastructure to be able to cope with the increased size of the ISP. A major finding is that HTTP is the dominant protocol used in today's access networks and has displaced Peer-to-Peer file sharing protocols. One reason for this is the popularity of file hosting sites, which emerged after the movie industry started pursuing users of P2P file sharing software. This trend is underlined by our flow statistics, showing some very large HTTP flows. The only still dominant P2P protocol is Bittorrent, with a share of about 38%, neglecting not classified flows. Flash video traffic is with 8% still the most widespread video container format, but we expect that a decrease with the discontinued development of mobile flash and the trend towards HTML5 video streaming.

Looking at the number of concurrent flows and flow inter-arrival times, network planners have to be aware of how to cope with it as the Internet usage increases drastically. Our results show that on average 22 GB of data is transferred per user per month compared to 12 GB per month measured only one year before. Class-based traffic shaping like used by our provider have to be able to handle this short flow inter-arrival times. For future research, we want to initiate new measurements, as the provider now offers IPTV, which results in a lot more traffic per user and a changed application distribution.

Acknowledgments. The authors would gratefully thank Antonio Pescapé from the University of Napoli and Phuoc Tran-Gia from the University of Wuerzburg for the fruitful discussions and support on this paper.

References

1. Pries, R., Wamser, F., Staehle, D., Heck, K., Tran-Gia, P.: Traffic Measurement and Analysis of a Broadband Wireless Internet Access. In: IEEE VTC Spring 2009, Barcelona, Spain (2009)
2. Maier, G., Feldmann, A., Paxson, V., Allman, M.: On Dominant Characteristics of Residential Broadband Internet Traffic. In: Proceedings of the 9th ACM SIGCOMM Conference on Internet Measurement Conference, IMC 2009, pp. 90–102. ACM, New York (2009)
3. Karagiannis, T., Papagiannaki, K., Faloutsos, M.: BLINC: Multilevel Traffic Classification in the Dark. SIGCOMM Comput. Commun. Rev. 35, 229–240 (2005)
4. Szabó, G., Szabó, I., Orincsay, D.: Accurate traffic classification. In: Proceedings of IEEE International Symposium on a World of Wireless Mobile and Multimedia Networks, WoW-MoM (2007)
5. Finamore, A., Mellia, M., Meo, M., Munafo, M.M., Rossi, D.: Experiences of Internet traffic monitoring with Tstat 25(3), 8–14 (2011)
6. García-Dorado, J., Finamore, A., Mellia, M., Meo, M., Munafò, M.M.: Characterization of ISP Traffic: Trends, User Habits and Access Technology Impact. Technical Report TR-091811, Telecommunication Networks Group - Politecnico di Torino (2011)
7. Cisco Systems: Regulating Packet Flow on a Per-Class Basis - Using Class-Based Traffic Shaping. Cisco IOS Quality of Service Solutions Configuration Guide (2010)
8. Dainotti, A., de Donato, W., Pescapé, A.: TIE: A Community-Oriented Traffic Classification Platform. In: Papadopouli, M., Owezarski, P., Pras, A. (eds.) TMA 2009. LNCS, vol. 5537, pp. 64–74. Springer, Heidelberg (2009)
9. Gamer, T., Mayer, C., Schöller, M., Gamer, T., Sciences, C.: PktAnon - A Generic Framework for Profile-based Traffic Anonymization. PIK Praxis der Informationsverarbeitung und Kommunikation 2, 67–81 (2008)
10. CAIDA: Preliminary measurement specification for internet routers (2004)
11. McKeown, N., Anderson, T., Balakrishnan, H., Parulkar, G., Peterson, L., Rexford, J., Shenker, S., Turner, J.: OpenFlow: Enabling Innovation in Campus Networks. SIGCOMM Comput. Commun. Rev. 38, 69–74 (2008)
12. Loiseau, P., Gonçalves, P., Primet Vicat-Blanc, P.: A Comparative Study of Different Heavy Tail Index Estimators of the Flow Size from Sampled Data. In: MetroGrid Workshop, Grid-Nets. ACM Press, New York (2007)
13. Economist Intelligence Unit and IBM Institute for Business Value: Digital economy rankings 2010 - Beyond e-readiness (2010)

From ISP Address Announcement Patterns to Routing Scalability

Liang Chen[1], Xingang Shi[2], and Dah Ming Chiu[1]

[1] Department of Information Engineering, The Chinese University of Hong Kong
{cl009,dmchiu}@ie.cuhk.edu.hk
[2] Network Research Center, Tsinghua University
shixg@cernet.edu.cn

Abstract. The Internet routing table size has been growing rapidly. In future Internet, if given a much larger public address space (e.g., IPv6), the potential expansion can be very significant, and the cost of a large routing table will affect many ISPs. Before devising methods to ensure routing scalability, it is necessary to understand what factors lead to the expansion of routing tables and to what extent they impact. In addition to the well known factors such as multi-homing, traffic engineering and non-contiguous address allocations, the tendency towards convenient address management also increases the routing table size. In this paper, we take a measurement-based approach to examine quantitatively how various factors, especially the current address aggregation status of different types of ISPs, contribute to the growth of global routing table. We show how these patterns affect the routing scalability, and discuss the implications as we plan address management and routing for the future Internet.

1 Introduction

The rapid growth of routing table size has become a concern for network operations long term. It is well known that there are already more than 370K IPv4 BGP entries (prefixes) in the Internet core routers. Once a larger address space, such as IPv6, is introduced in the Internet, this scalability issue will become more serious. Before devising methods to ensure routing scalability, we need to understand what leads to the expansion of routing tables and to what extent they contribute to the scalability issue quantitatively.

The Internet is composed of tens of thousands of Autonomous Systems (ASes). An AS can be loosely considered as being an Internet Service Provider (ISP), although an ISP may own multiple ASes, and some ASes are operated by an organization rather than a business. Since each AS has its own operation and management strategies, the address announcement behavior of one type AS is often different from another.

ISP address announcement provides a clue to reveal the essence of routing scalability issue quantitatively. In this paper, we take a close look at the prefix announcement patterns, and explore the current status of address usage for

A. Pescapè, L. Salgarelli, and X. Dimitropoulos (Eds.): TMA 2012, LNCS 7189, pp. 43–47, 2012.
© Springer-Verlag Berlin Heidelberg 2012

different kinds of ISPs. We propose a measurement metric, the *address aggregation ratio*, to reflect the extent each type of ISP contributes to the scalability issue. We also point out four thresholds in routing table size based on our measurement results of address usage. Besides some well-know facts, our study indicates that the address management for convenience is actually another factor contributing to the growth of routing table.

There exist several works in understanding the routing scalability problem qualitatively. The common policy of allocating addresses as needed by customers leads to address fragmentation. Incremental address provisioning, as well as multi-homing and traffic engineering, have been driving the alarming growth of the Default Free Zone (DFZ) routing table size [5]. The previous works [2,4] study several factors contributing to the routing scalability issue with effective approaches. Other approaches proposed for this problem include separating edge networks from transit networks, address aggregation based on geographic information[6], and aggregation of FIB entries [7]. Our work extracts common patterns of current ISPs behavior with in-depth study on address announcements, to understand what should be considered when designing a practical solution.

2   Methodology and Data Source

We collect historical data from Internet Registries and RouteViews to study ISP behavior in address allocation and announcement. We take a snapshot record of global routing table in August 2011 and consider its evolution in the past decade as needed. Each prefix in the global routing tables represents a specific block of addresses and how it can be reached from the whole Internet.

We collect a complete list of allocated address blocks from RIRs, and investigate how they are announced by different ISPs. The second piece of data is collected from the RouteViews project, which collects BGP routes from many ASes in the core of the Internet, so that it provides a global view of the Internet routing system. We study not only the spatial and temporal patterns of address announcement, but also how the announcement deviates from the original allocation. At last, we manually classify the ISPs which contribute the most prefix announcement in the global routing table, and study the address aggregation status of different types of ISPs.

3   Address Announcement Patterns and ISP Behavior

By observing the address announcements, we can understand the current usage of IPv4 addresses in Internet. Our investigation indicates that: a) Multi-homing contributes nearly 25% more prefixes; b) Traffic engineering introduces the most part of additional prefixes; c) Non-contiguous address allocations expand the base of announced addresses; d) Convenient address management accounts for a certain proportion of the additional prefixes. These findings are also supported by several previous works, and due to space limitation, we omit the details.

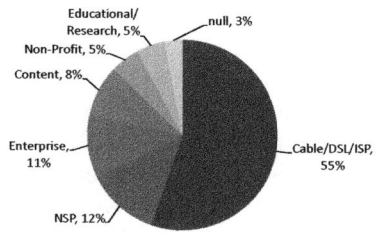

Fig. 1. Classification for Top ~2000 ASes

Table 1. Percentage of /24 Prefixes and Aggregation Ratio for Top ~2000 ASes

Type	/24 (%)	Aggregation ratio r
Content ISPs	57.94	12.25
Access ISPs	49.57	12.77
Backbone ISPs	51.01	13.93

We suspect the reason for the number of prefixes announced, and the aggregation efficiency of an ISP may depend on the type of its business. Our intuitive understanding is that there are at least three types of ISPs: (a) Access ISPs which typically provide residential cable or DSL access, (b) Backbone ISPs which provide services for small ISPs, and (c) Content ISPs which often provide content hosting or distribution. We tried to find a public source where ISPs are categorized according to these loose type of categories, but in vain. One earlier work [3] classified ISPs into large ISPs, small ISPs, customer networks, universities, Internet exchange points and network information center. However many new ASes, especially ASes focused on providing content distribution, have emerged in the past few years. Another public source [1] categorized ISPs into the following six categories: Cable/DSL/ISP, NSP, Content, Non-Profit, Enterprise, and Education/Research, which is close enough to our needs. But their classification covered only 22.1% of the top ~2000 ASes who contribute around 2/3 of prefixes in routing tables. So we decided to manually classify the rest of these top ~2000 ASes, based on their name, and other public information gathered using search engine. The breakdown of our classification is shown in Fig. 1. Obviously, Cable/DSL/ISPs and NSPs account for the majority of these ASes.

To investigate the aggregation efficiency for each category of ISPs, we use n to represent the number of prefixes announced by an ISP, use s to represent the number of addresses covered by these prefixes, and use a metric r to measure the *address aggregation ratio*, which is defined as $r = \log_2 \left(\frac{s}{n} \right)$. Actually, r is the average number of addresses covered by each prefix (in a log manner), and a larger r indicates better aggregation efficiency.

Based on the results shown in Table 1,[1] the address aggregation of Backbone ISPs is more efficient than that of Access or Content ISPs. Since NSPs usually allocate addresses to end users hierarchically, they tend to aggregate addresses by their management. It is insightful that Content-type organizations favor less aggregated prefixes to route their traffic in average. It is likely that the emergence of more Content or Access ISPs (the periphery of the Internet) will further burden the Internet routing system. Also, it is worth noting that only Backbone ISPs have a larger address aggregation ratio than the average result of the whole Internet.

---

[1] In Table 1, to be more concise, we group Cable/DSL/ISP, Non-Profit, Enterprise and Education/Research together as Access ISPs, Backbone ISPs are NSPs.

Table 1 also lists the percentage of /24 prefixes among all prefixes announced by each category of ASes. It is worth noting that Content ISPs tend to use /24 prefixes more frequently than Access or Backbone ISPs. While currently Access ISPs announce much more /24 prefixes than the other two, they do not use /24 prefixes as aggressively as Content ISPs.

Hierarchical routing based on aggressive address aggregation could be one solution to handle the routing scalability issue. Since ASes are the entities announcing prefixes in global Internet, the number of ASes would be the best case for the size of routing table in an ideal hierarchical routing system. Once each AS is allocated multiple address blocks, the optimal address aggregation would be in the number of allocations. However, the AS usually divides the allocated blocks into multiple prefixes to utilize the address space, which can be indicated by the number of aggregated prefixes in our study. In the worst case, when optimization of network operation such as multi-homing or traffic engineering is involved and supported by the routing system, the routing table size will be the number of total prefixes as in current practice.

Therefore, the number of ASes, the number of allocated address blocks, the number of aggregated prefixes and the number of total prefixes could be four possible thresholds for the size of global routing table. Each case trades off certain practical factors in both address allocation and announcement.

If we can monitor each type of ISP in terms of their usage of global routing table, it is possible to provide one approach to alleviate the routing scalability issue. When every AS's usage of the global routing resource is publicly monitored, the pressure on those ones who contribute most may help save their announcement of unnecessary prefixes. Also the efficiency of the routing system may be improved. As a trivial suggestion, this could be considered in both current and future Internet when concerning the routing scalability issue.

4   Conclusions

We believe there is much to be learned from the way addresses are allocated and announced in the current Internet, and the insights can be useful for routing scalability studies for future Internet. In this paper, we have taken a quantitative approach to analyze current address allocation and announcement practices, and draw a number of conclusions about announcement patterns and ISP behaviors on address aggregation. For future directions, we hope to use the statistics we collect and the metric we proposed to help to evaluate the scalability issue for routing system, and provide further in-depth analysis.

References

1. Networktools (July 2011), http://www.networktools.nl/
2. Bu, T., Gao, L., Towsley, D.: On characterizing BGP routing table growth. Computer Networks 45(1), 45–54 (2004)

3. Dimitropoulos, X., Krioukov, D., Riley, G.: Revealing the autonomous system taxonomy: The machine learning approach. In: PAM Workshop (2006)
4. Meng, X., Xu, Z., Zhang, B., Huston, G., Lu, S., Zhang, L.: IPv4 address allocation and the BGP routing table evolution. ACM SIGCOMM CCR 35(1), 71–80 (2005)
5. Meyer, D., Zhang, L., Fall, K.: Report from the IAB Workshop on Routing and Addressing. RFC 4984, IETF (September 2007)
6. Yin, X., Wu, X., Chon, J., Wang, Z.: ISPSG: Internet Service Provider-Separated Geographic-Based Addressing and Routing. In: IEEE GLOBECOM Workshops (2009)
7. Zhao, X., Liu, Y., Wang, L., Zhang, B.: On the Aggregatability of Router Forwarding Tables. In: IEEE INFOCOM (2010)

I2P's Usage Characterization

Juan Pablo Timpanaro[1], Isabelle Chrisment[2], and Olivier Festor[1]

[1] INRIA Nancy-Grand Est, France
[2] LORIA - ESIAL, Henri Poincaré University, Nancy 1, France

Abstract. We present the first monitoring study aiming to characterize the usage of the I2P network, a low-latency anonymous network based on garlic routing. We design a distributed monitoring architecture for the I2P network and show through three one-week measurement experiments the ability of the system to identify a significant number of all running applications, among web servers and file-sharing clients.

1 Introduction

The I2P network[1], mainly designed to allow a fully anonymous conversation between two parties inside the network, contains a full range of available applications. The goal of this work is to characterize the usage of the I2P network and thereby to answer the following question: what is the I2P network used for?. We consider that this analysis is important for the network improvements. By determining that most of I2P users are, for example, file-sharers, we could enhance the network in that direction.

To do that, we propose a fully operational architecture to monitor the I2P network at application-level. Then, we show we can identify which applications are most used and when they are used, as opposed to statistics web sites[2] giving only the number of applications, but not the type.

2 I2P Monitoring Architecture

In the I2P network, a distributed hash table based on the Kademlia [1] protocol is used to store and share network metadata. However, contrary to the Kademlia protocol, only the fast I2P users form part of the DHT. They are called the *floodfill* peers. There are two types of network metadata: *leasesets* and *router infos*. A leaseset provides information about a specific destination, like a web server, a BitTorrent client, an e-mail server,.... A router info provides information about a specific router and how to contact it, including the router identity (keys and a certificate), the address to contact it (ip and port) and a signature.

---

[1] http://www.i2p2.de/
[2] Such as http://stats.i2p.to/

A. Pescapè, L. Salgarelli, and X. Dimitropoulos (Eds.): TMA 2012, LNCS 7189, pp. 48–51, 2012.
© Springer-Verlag Berlin Heidelberg 2012

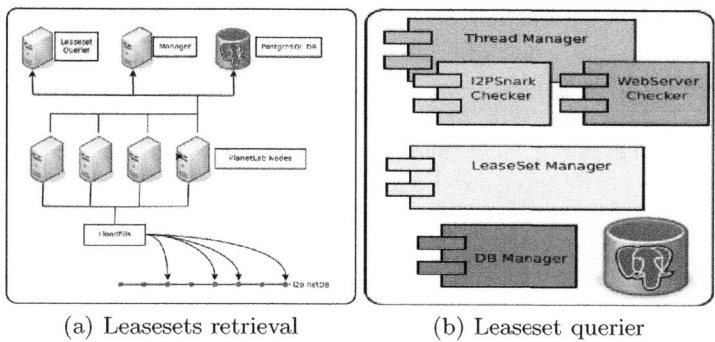

(a) Leasesets retrieval   (b) Leaseset querier

Fig. 1. Complete monitoring architecture

Our monitoring architecture is divided in two main parts:

– A first part is responsible for retrieving and collecting *leasesets* as described
  in Figure 1(a). Each floodfill is running in one PlanetLab node, and logs
  every received request to a PostgreSQL server, located in our security lab[3].
– A second part, the *leaseset querier*, runs in parallel and is shown in Figure
  1(b). The *Leaseset Manager* periodically retrieves new leasesets from the
  database, while the *Thread Manager* creates different threads to test a given
  destination for a particular application (a Web Server or an I2PSnark client).
  Once a leaseset is tested, the result is kept in the PostgreSQL server.

We focus mainly our study on two kinds of applications running on top of an I2P
router and published in the DHT: a web server and an I2PSnark client, which
is a modified BitTorrent client.

To tag a given leaseset as a web server, we open an I2P connexion through
it and send a GET message. If the response contains well-known http keywords,
then that leaseset corresponds to a web server.

To test an I2PSnark client, we use the following approach. Once the I2P
connection is established, we send a first message, a well-formed BitTorrent
message, requesting a random torrent. If that given leaseset is actually running
an I2PSnark client and not sharing the torrent, it will immediately close the
socket. Secondly, we re-open the connection and send a malformed BitTorrent
message. If the response timeouts, then we conclude that the given leaseset is
running an I2PSnark client. If we do not receive any answer for the first message,
we can not assume anything.

3   Experiments

3.1   Setup

During these first series of experiments we used the Planet Lab test-bed. Planet
Lab nodes have restrictions regarding the available bandwidth, and therefore

---

[3] http://www.inria.fr/actualite/mediacenter/inauguration-lhs-nancy

we placed 15 floodfills with minimum bandwidth settings, still allowing us to perform as floodfills nodes. Our leaseset querier ran in parallel, analysing new leasesets and storing the results. We conducted 3 different one-week experiments, starting on September 12th, September 27th and October 4th. We logged four possible outcomes for a given leaseset: *WebServer, I2PSnark, Unknown* (The given leaseset is not running a web server nor an I2PSnark client) and *Destination not found* (The given leaseset is unreachable).

3.2 First Results

Figure 2 gives the percentage of leasesets that we were able to identify and we tagged for the first experiment. In average, we identified 32.06% of all the leasesets queried, with most of them being I2PSnark clients, as shown in figure 2(b).

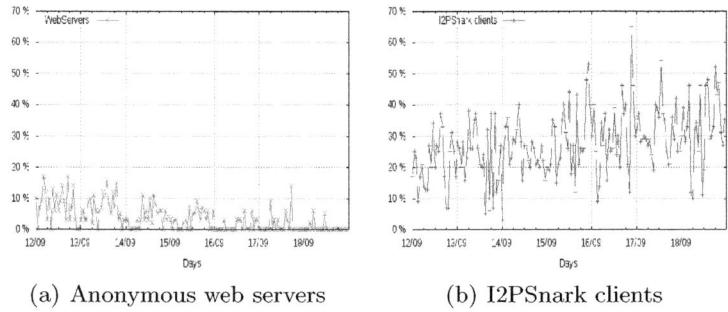

(a) Anonymous web servers   (b) I2PSnark clients

Fig. 2. Identifiable leasesets (September 12th - 19th)

Table 3 summarizes the number of I2P destinations we analyzed for every experiment. For our last experiment, the *Destination not found* value is higher than we saw in the previous experiments. The statistics web page of I2P[4] reports that between the 10/02 and the 10/09, the number of routers in the network increased considerably (from 7000 routes to 10000 routers.) This increase of new participants in the network could explain why we have a higher number of unreachable destinations, since many I2P users might have been trying the network and testing different applications.

	Anonymous web servers	I2PSnark clients	Unknown	Destination not found	Total
First experiment	193	1312	1423	1793	4721
Second experiment	172	1106	1503	1757	4538
Third experiment	175	1057	1186	2349	4767

Fig. 3. Results for every one-week experiment

We have, on average, 30.16% of *unknown* applications, which means that we successfully opened a connexion through these leasesets, but we failed at tagging them.

---

[4] http://stats.i2p.to/

4   Related Work

McCoy et al.[2] analysed the application layer of outgoing traffic in Tor, to determinate the protocol distribution in the network. This study resembles what we want to achieve for I2P, however, the data collection methodology applied by McCoy et al. can not be used in I2P. More recently, Loesing et al. [3] presented a measurement of sensitive data on the Tor network, such as countries of connections and outbound traffic by port. Additionally, Loesing [4] measured the *trends* of the Tor network from directory information. Nevertheless, this network has a central component, a *directory server* and hence the monitoring approaches can not be applied to the I2P network.

5   Conclusion

This study shows a series of experiments on the I2P network. After the first one-week experiment, and after analysing 4721 leasesets, we were able to identify 193 web servers and 1312 I2PSnark clients, and we determined that 37% of the published leasesets were off-line after their publication on the netDB.

Monitoring a network is crucial to understand what it is used for. We consider it a mistake to apply previous well-known results in p2p networks, like most users performing file-sharing during weekends, in anonymous networks before classifying its traffic accordingly. Anonymous networks allow a user to keep its identity safe while web-surfing or file-sharing, for example. Moreover, an anonymous network like I2P, which provides a set of built-in services, can be used for different objectives and in different manners when comparing it to non-anonymous networks, or even to anonymous networks, like *Tor*.

This work is the first step on monitoring and classifying the traffic on the I2P network, in order to understand the network. A full version of this work can be found in [5]. Other experiments and measurements are in progress.

Acknowledgment. We thank the anonymous leading developer of I2P for his useful comments and reviews of this paper.

References

1. Maymounkov, P., Mazières, D.: Kademlia: A Peer-to-Peer Information System based on the XOR Metric. In: Druschel, P., Kaashoek, M.F., Rowstron, A. (eds.) IPTPS 2002. LNCS, vol. 2429, pp. 53–65. Springer, Heidelberg (2002)
2. McCoy, D., Bauer, K., Grunwald, D., Kohno, T., Sicker, D.: Shining Light in Dark Places: Understanding the Tor Network. In: Borisov, N., Goldberg, I. (eds.) PETS 2008. LNCS, vol. 5134, pp. 63–76. Springer, Heidelberg (2008)
3. Loesing, K., Murdoch, S.J., Dingledine, R.: A Case Study on Measuring Statistical Data in the TOR Anonymity Network. In: Sion, R., Curtmola, R., Dietrich, S., Kiayias, A., Miret, J.M., Sako, K., Sebé, F. (eds.) FC 2010 Workshops. LNCS, vol. 6054, pp. 203–215. Springer, Heidelberg (2010)
4. Loesing, K.: Measuring the tor network from public directory information. In: Proc. HotPETS, Seattle, WA (August 2009)
5. Timpanaro, J.P., Isabelle, C., Olivier, F.: Monitoring the I2P network. Research Report RR-7844, INRIA (October 2011)

Experimental Assessment of BitTorrent Completion Time in Heterogeneous TCP/uTP Swarms

Claudio Testa[1], Dario Rossi[1], Ashwin Rao[2], and Arnaud Legout[2]

[1] Telecom ParisTech, Paris, France
{first.last}@enst.fr
[2] INRIA Planete, Sophia Antipolis, France
{first.last}@inria.fr

Abstract. BitTorrent, one of the most widespread used P2P application for file-sharing, recently got rid of TCP by introducing an application-level congestion control protocol named uTP. The aim of this new protocol is to efficiently use the available link capacity, while minimizing its interference with the rest of user traffic (e.g., Web, VoIP and gaming) sharing the same access bottleneck.

In this paper we perform an experimental study of the impact of uTP on the torrent completion time, the metric that better captures the user experience. We run BitTorrent applications in a flash crowd scenario over a dedicated cluster platform, under both homogeneous and heterogeneous swarm population. Experiments show that an all-uTP swarms have shorter torrent download time with respect to all-TCP swarms. Interestingly, at the same time, we observe that even shorter completion times can be achieved under mixtures of TCP and uTP traffic, as in the default BitTorrent settings.

1 Introduction

Though some might argue that network congestion control is a problem that has been studied to death–to which we tend to agree, at least concerning the large amount of literature on the topic–yet the network architecture and usage are undergoing profound changes, that make the study of congestion control issues once more necessary.

As far as the architectures are concerned, recent research has, e.g., addressed the TCP incast problem in data center networks. As far as the usage is concerned, we have lately witnessed to an explosion of new application-layer flow and congestion control algorithms [1–3], which are usually implemented at the application-layer over either TCP [1] or UDP [2, 3]. Depending on the application they have been built for, these protocols have rather different goals that deeply influence their design.

In this work, we focus on the BitTorrent filesharing protocol that recently replaced TCP by uTP[1] [4, 5], a lower than best effort protocol for data transport on top of UDP. uTP starts from the observation that nowadays the Internet bottleneck is typically at the user ADSL access link: hence, congestion typically happens between different flows of the same user. Moreover, since ADSL modems have rather long buffers (up to few seconds [6, 7]), using TCP for non-interactive but massive data downloads has a possibly negative impact on interactive communication (e.g., Skype, gaming, etc.). In

---

[1] Notice that this new protocol has two names: uTP in the BitTorrent community [4], and LED-BAT in the IETF community [5], that we may use interchangeably in the following.

A. Pescapè, L. Salgarelli, and X. Dimitropoulos (Eds.): TMA 2012, LNCS 7189, pp. 52–65, 2012.
© Springer-Verlag Berlin Heidelberg 2012

other words, as TCP fills the buffer, the self-induced congestion translates into high latency and losses, that possibly harms other interactive application. To avoid the troubles of self-induced congestion and at the same time be efficient for massive data download, uTP tries to limit the end-to-end delay (by reaching a fixed amount of *target delay* in the access buffer) while maximizing the utilization of the access capacity. As the queuing delay is bounded with uTP, this improves users experience for interactive applications.

While uTP has sound and appealing goals, it is clearly understood that users will be the ultimate judge of BitTorrent performance, as in BitTorrent's own words *"unless we can offer the same performance [of TCP], then people will switch to a different BitTorrent client"* [8]. Our recent work [9] suggests that uTP performs better than standard TCP, as the use of uTP practically limits the queuing delay to a small target, this translates into faster signaling as a side effect. However, results in [9] are based on ns2 simulation: it becomes thus imperative to assess whether the observed phenomena also happens in practice, which is precisely the scope of this work.

We run BitTorrent applications in a flash crowd scenario over the Grid'5000 platform [10], with special attention to the main user-centric metric, the torrent completion time. Results of our experiments confirm our previous simulation results, in that, as observed in [9], uTP can reduce the torrent download time.

Yet, this experimental work brings new insights beyond [9]. Currently, the default settings of BitTorrent yield to the use of a *mixture of TCP and uTP traffic*. Hence, in this work we evaluate how this choice performs compared to the cases in which all peers use only a single protocol between TCP and uTP. In this case, our experimental results show that completion time under heterogeneous swarms can be even lower than all uTP (and, clearly, all TCP) swarms.

The remainder of this paper is organized as follows. First, Section 2 discusses the related work. Section 3 reports preliminary insights on low-level BitTorrent settings gathered from a small local testbed, which are instrumental to our experiments, whose results are reported in Section 4. Finally, we summarize the paper and discuss future work in Section 5.

2   Related Work

Two bodies of work are related to this study: on the one hand, we have work focusing on congestion control aspects [1, 2, 11–13], and on the other hand work focusing on Bit-Torrent [7, 9, 14–20]. First, congestion control literature already proposes several protocols aiming, as LEDBAT, to achieve lower-than-TCP priority, of which TCP-LP [12], NICE [11], 4CP [13] are notable examples. Yet, we pinpoint a recent tendency toward moving congestion and flow control algorithms *from the transport layer to the application layer,* of which uTP [3] for background file-sharing, Skype [2] for interactive VoIP and YouTube [1] for interactive VoD are again notable examples. Unlike transport-layer congestion control, that applies to classes of applications, these application-layer congestion control protocols are usually built for single applications, with specific requirements in mind: these "one of a kind" deployments will in our opinion need further attention in the future. Recently LEDBAT also got the attention of Apple's developers, resulting in a implementation for MAC OS (preliminary tests available at [21]).

Second, BitTorrent literature dissected many aspects of this successful P2P protocol, from the pioneering time of [14]. While our own previous works, such as [15, 16], already study BitTorrent download performance by means of either passive measurements or experimental tests (as in this work), however they report on performance at a time when BitTorrent was using TCP, and should thus be updated in light of BitTorrent recent evolution. More generally, though related work on uTP exists [7, 9, 17, 17–20, 22]), it does however adopt a congestion control perspective (with the exception of [9]). In particular, an experimental approach is adopted in [6, 7, 17]: [17] attacks the problem of clock drift in uTP, while [7] performs a black box study of initial proprietary versions of the protocol and [6] focuses on the interaction of uTP and active queue management techniques that are becoming commonplace in modern home gateways. A simulative approach is instead adopted in [18–20, 22]: a fairness issue of uTP is revealed in [18] and solved in [19], while [20] compares (i) the level of low-priority of TCP-LP [12], NICE [11] and uTP [3] and (ii) the fairness of TCP and uTP, and finally [22] investigates policies for dynamic parameter tuning.

The only previous work addressing the impact of uTP on BitTorrent completion time is our own recent work [9], that however employs ns2 simulations unlike in this work. Interestingly, some of the observations of this study are in agreement with [9], e.g., showing a larger completion time for increasing buffer occupancy on the data plane. Yet, we point out that [9] does not consider an hardcoded preference for uTP, nor bidirectional uTP connections: hence, an interesting difference with respect to the current work is that [9] forecasted heterogeneous performance for heterogeneous swarms (i.e., larger completion time for TCP peers) that we have shown not to hold on practice.

3 Preliminary Insights

As previously stated, the uTP protocol aims at jointly (i) being efficient by fully exploiting the link capacity when no other traffic is present, and (ii) being low priority by yielding to other competing traffic on the same bottleneck. In order to achieve both these goals, uTP needs to insert only a limited amount of packets in the bottleneck buffer: on the one hand, since the queue is non empty, the capacity is fully exploited. On the other hand, as the queuing delay in the buffer remains bounded, this does not harm interactive applications.

uTP exploits the ongoing data transfer to measure the One-Way Delay (OWD) on the forward path. While measuring the OWD is notoriously tricky among non-synchronized Internet hosts, uTP is interested in the *difference* between the current OWD and the minimum OWD ever observed (used as an approximate reference of the base propagation delay). In turn, this OWD difference yield to a measure of the current *queuing delay*, that is used to drive the congestion window dynamics: when the measured queuing delay is below a given *target delay*, the congestion window grows, but when the queuing delay exceeds the target the congestion window shrinks.

The impact of this new protocol on the performance of BitTorrent can be affected by essentially two different settings. At a single flow level, uTP is primarily driven by the uTP *target delay* setting. At a swarm level, peers *relative preference* for TCP vs uTP protocols plays an important role as well. Hence, before running a full-fledged set of experiments, we need to get some preliminary insights on the settings of the above two parameters. In more details, these are: (i) *net.utp_target_delay*, that tunes the value of uTP delay of each flow, and (ii) *bt.transp_disposition*, that drives TCP vs uTP preference of the BitTorrent client.

To this aim, we perform a battery of tests in a completely controlled environment involving one seed and two leechers, all running the latest version of the uTorrent client available for Linux (3.0 build 25053, released on March 2011). Clients in the local testbed are interconnected by a 100 Mbps LAN and we use the Hierarchical Token Bucket (HTB) of `netem` to emulate an access bottleneck on the PC running the seed, whose uplink capacity is then capped at 5 Mbps. Any additional delay was added on the testbed links. The tracker is private within the testbed and is used to announce a set of three different torrent files having different file and chunk sizes (file size of 10, 50 and 100 MBytes, and chunk size of 256, 512 and 1024 KBytes respectively).

To understand the BitTorrent settings, we tweaked the default configuration[2] aiming at (i) verifying the compliance of *net.utp_target_delay* to the imposed delay and (ii) understanding how *bt.transp_disposition* settings, which controls when uTP is used, impacts the performance of BitTorrent.

In the case of (i) *net.utp_target_delay*, usually a single experiment is sufficient to verify its compliance, since every flow obeys its own setting. Conversely multiple experiments were necessary in the case of (ii) *bt.transp_disposition*, since the behavior of a peer is affected by the *bt.transp_disposition* value of other peers as well.

In the following we summarize the most relevant findings of the local testbed experiments. Overall, in these tests we captured about 2 GBytes of packet level traces, that we make available to the scientific community at [23]. The remainder of the experiments, performed on the Grid'5000 platform, are reported in Section 4.

3.1   *net.utp_target_delay*: Target Delay Settings

The *net.utp_target_delay* parameter stores the value of the uTP target delay in milliseconds, and its default value is equal to 100 ms as stated in the BEP29 [4]. Furthermore, uTP is also standardized at the IETF LEDBAT Working Group [3], which specifies 100 ms as a *mandatory upper bound value* (while earlier version of the draft referred to a 25 ms delay target).

Yet, the GUI of the Windows client allows to modify its default value, opening the way for competition between legacy applications. This behavior is confirmed by Fig. 1, where we show two experiments, performed at different times, where a single uTP flow sends data on a 5 Mbps bottleneck, with different values of the uTP target delay. We see in Fig. 1 that for both target delay settings, BitTorrent is using the entire available capacity, and the end-to-end delay corresponds to the target that we set by means of the *net.utp_target_delay* parameter.

---

[2] uTorrent clients store their configuration in a file which is not directly editable (as it also contains an hash value on the configuration content, performed by the client itself) and moreover Linux GUIs do not offer the possibility of modifying the default settings. However, the configuration file format used by Linux clients is the same as the one of the Windows clients: hence, we used Windows GUIs to produce a pool of configuration files, that we later loaded in Linux.

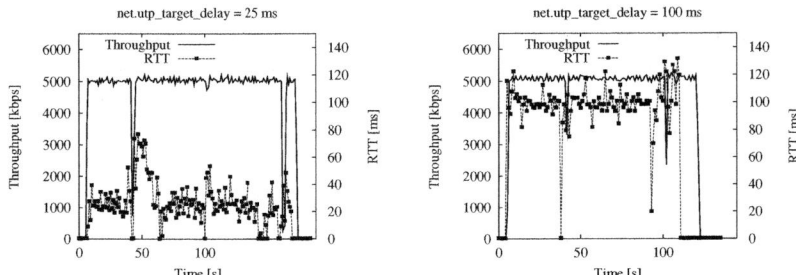

Fig. 1. uTP Target Settings: RTT and throughput for a single flow with *net.utp_target_delay*=25 ms (left) and *net.utp_target_delay*=100 ms (right)

As the uTP/LEDBAT specifications [3, 4] refer to a mandatory target value, we comply to the standard and focus in the remainder of this paper on the study of swarms with the same default value for *net.utp_target_delay*. At the same time, we point out that as a future work, it would be interesting to see whether, by tweaking the *net.utp_target_delay* value, some peers (or some applications) can gather an advantage over the rest of the swarm.

3.2 *bt.transp_disposition*: TCP vs uTP Settings

The second parameter, namely *bt.transp_disposition*, controls which protocol is used for the incoming and outgoing data connection of the client. As reported in the online uTorrent manual[3], *bt.transp_disposition* is a bitmask that sums up the following behaviors:

- 1: attempt outgoing TCP connections
- 2: attempt outgoing uTP connections
- 4: accept incoming TCP connections
- 8: accept incoming uTP connections
- 16: use the new uTP header format

uTorrent default value is 31, which means that the client will accept both TCP and uTP flavors, for either sending or receiving data, possibly using the new uTP header format.

To understand the implications of *bt.transp_disposition* settings, we perform a number of tests with heterogeneous settings of the client. Notice that the parameter space we explore and that we make available at [23] is larger than the one reported in Tab. 1. Yet, for the sake of simplicity, we only report in the table the cases that we later study in Section 4, which already conveys some interesting information. Notice also that in all the experiments, the seed is set with the default value *bt.transp_disposition*=31.

In *Case 1*, the two leechers A and B have different setting for the *bt.transp_disposition* parameter: more precisely, A should attempt data connection only using TCP while B should use uTP (and both will accept every flavor in reception). Our experiments show that in this scenario, peer B sends data to peer A using the uTP protocol, which is the expected behavior. However, peer A sends data to peer B using the uTP protocol too,

---

[3] *http://www.bittorrent.com/help/manual/appendixa0212#bt.transp_disposition*

Table 1. TCP vs uTP BitTorrent transport disposition

case	peer	attempt	accept	disp	B → A	A → B	comment
1	A	TCP	*	13	uTP	uTP	Hardcoded uTP preference
	B	uTP	*	14			
2	A	TCP	TCP	5	TCP	TCP	Legacy BitTorrent implementations
	B	*	*	31			

which happens consistently over all repetitions, and irrespectively of file and chunks size. The reason is that when a uTP connection from B to A is opened, peer A can use this opened bidirectional connection to send data to peer B. Besides, as confirmed by Arvid Norberg, one of the main BitTorrent developers, the uTorrent client has a *hard-coded uTP preference*, so that in case both a TCP and a uTP connection will be successfully established, the former will be closed and only the latter will be used. As we will see in Section 4, this preference has some important (and beneficial) consequences on the overall swarm completion time.

In *Case 2*, leecher A attempts and accepts only connection via TCP (as a legacy BitTorrent implementation would do), while leecher B maintains the default value for *bt.transp_disposition* (which means to attempt and accept both protocols). In this scenario, any communication between the two peers are performed using the TCP protocol, which is consistent and expected for backward compatibility with older BitTorrent clients.

Other cases, not shown in Tab. 1, yield to different shares of traffic between TCP and uTP. At the same time, since the number of leechers is small, the exact value of the breakdown is heavily influenced by the seeder flavor as well. As such, we defer a quantitative analysis of such a breakdown in the next section.

4 Experimental Results

We now report the experimental results on the impact of uTP and TCP transport on the torrent completion time. First we briefly describe the Grid'5000 experimental platform and then focus on two case studies, namely (i) homogeneous and (ii) heterogeneous swarms, depending on the *bt.transp_disposition* settings for the leechers.

Homogeneous settings refer to scenarios were all peers have either a TCP-only preference (*bt.transp_disposition*=5, which mimic the behavior of old uTorrent versions or legacy applications), or a uTP-only preference (*bt.transp_disposition*=10, in case uTP will prevail over TCP), or are able to speak both protocols (*bt.transp_disposition*=31, the current default behavior, though with an hardcoded preference for uTP as we have seen in Section 3.2).

Homogeneous settings provide a useful reference, but we must consider also experiments with heterogeneous scenarios that correspond to what is observed in the Internet with clients that do not support uTP at all, or that support uTP but as a fallback choice rather than the default one.

We therefore investigate heterogeneous settings as well, considering scenarios with different ratios of peers using uTP and TCP. More precisely, we consider the case where peers are able to accept any incoming protocol, but have different preferences for the uplink protocol (*bt.transp_disposition*=13 for TCP, and *bt.transp_disposition*=14 for uTP). We consider the case where the preference splits are 50/50, 25/75, or 75/25 to mimic scenarios where TCP vs uTP preferences are fairly balanced, or biased toward one of the two protocols.

Notice that, while there may be several uTP implementations available, different BitTorrent applications use different default settings (i.e., sticking to TCP preference or embracing uTP) depending on the success of the new protocol (and the existence of readily available libraries for different operating systems).

4.1 Grid'5000 Setup

We performed experiments on a dedicated cluster of machines that run Linux as the host operating system and using the uTorrent 3.0 client as before. Hosts of the Grid'5000 platform are interconnected by an high-speed 1 Gbps LAN, and we emulate realistic bandwidth restrictions and queueing of home gateways by using the `netem` module for the Linux kernel. As noted in [3] and experimentally confirmed by [6, 7], ADSL modems can buffer up to a few seconds worth of traffic: in our experiments, we set the buffer size B according to the uplink capacity C so that $B/C = 1$ second worth of traffic.

We instrumented the Linux kernel to log the queue size Q in bytes and packets after each dequeue operation, logging also the cumulative number of packets served and dropped by the queue. During our experiments, we disabled the large segment offloading [24] which ensured that the maximum segment size of the TCP and uTP packets never exceeded the maximum transfer unit (MTU). In each experiment we used the Cubic flavor of TCP, the default for Linux kernels: in reason of our previous work [7], we may expect Cubic to be more aggressive with respect to the standard TCP NewReno flavor, and more similar to the default TCP Compound flavor adopted in recent versions of Windows.

We use 76 machines on the Grid'5000 platform and consider an Internet flash crowd scenario, where a single seed is initially providing all the content (a 100 MBytes file) to a number of leechers all arriving at the same time (and never leaving the system). Furthermore, each BitTorrent peer (i) experiences an ADSL access bottleneck [25] and (ii) encounters a self-induced congestion, unrelated to the cross-traffic [3].

As for (i), we start by considering 3 simple homogeneous capacity scenarios in which we limit the leechers and seed uplink capacity to $C = 1$, 2 and 5 Mbps with the Hierarchical Token Bucket algorithm. For the sake of simplicity, as the qualitative results do not change for different values of C, in the following we report the results for $C = 1$ Mbps. While it could be objected that Internet capacity are not homogeneous, we argue that homogeneous scenarios are needed as a first necessary step before more complex and realistic environments are emulated. Additionally, the impact of heterogeneous access capacity is a well known clustering effect [16], that we believe to fall outside of our main aim, i.e., the comparison of TCP vs uTP transport, and that can be studied with a future experimental campaign.

As for (ii), we are forced to map a single peer on each host of the Grid'5000 platform, as otherwise unwanted mutual influence may take place on multiple peers running on the same hosts. Given the number of hosts $N = 76$ we can use, this constrains us on the size of the swarm we investigate. However, we prefer to take a cautious approach, and avoid to introduce the aforementioned mutual influence that could bias in an unpredictable way the results of our experiments (see a discussion on the conclusions).

We repeat the experiments three times for each settings, to smooth out stochastic variability in the experiments due to BitTorrent random decisions (e.g., chunk selection, choke, optimistic unchoking etc). Also, we make results of our campaign available to the scientific community as for the local testbed at [23]. Overall, the volume of collected data in the Grid'5000 testbed amounts to about 10 GBytes. Yet, we point out that, in reason of the large number of experiments and seeds, we were not able to store packet-level traces, but only periodic transport-layer (i.e., TCP, UDP traffic amount), application layer (i.e., tracker) and queuing logs.

4.2 Homogeneous *bt.transp_disposition* Settings

Results of the homogeneous case are reported in Fig. 2. For each metric of interest, the figure reports the *envelope* of the gathered results, i.e., the minimum and maximum curves over the three iterations.

We express results in terms of (i) the cumulative distribution function (CDF) of the torrent completion time T, on the right plots and of (ii) the complementary cumulative distribution function (CCDF) of the buffer occupancy Q of the access link of each peer, on the left plots. The buffer occupancy is expressed both in bytes (bottom x-axis) and in terms of the amount of delay an interactive application would experience for the emulated access capacity (top x-axis).

Additional details are reported in the inset of each figure, showing: (iii) the average $E[T]$ and standard deviation $\sigma[T]$ of the torrent completion time; (iv) the byte-wise share between TCP and uTP, with a notation X/Y that specifies that $X\%$ ($Y\%$) of the bytes are carried over TCP (uTP), with $X + Y = 100\%$; (v) the mean queue size $E[Q]$ in KBytes and milliseconds.

In the top row we report the scenario where all peers use default settings (*bt.transp_disposition*=31), i.e., the peers are able to speak both TCP and uTP protocols. Middle plots report the case of an all-uTP swarm (*bt.transp_disposition*=10), while all-TCP swarms (*bt.transp_disposition*=5) are depicted in the bottom row.

Notice that, on the long run, all swarms achieves similar efficiency: looking at the CDF of the buffer occupancy in Fig. 2 we can see that in roughly 80% of the time, after a dequeue operation the queue is non-empty. That efficiency is also tied to the BitTorrent system dynamics (e.g., pipelining of the requests, chunk exchange dynamics, etc.). Also the number of packets remaining in the queue after a packet transmission further depends on the congestion control protocol of choice. As expected, TCP AIMD dynamics tend to fill the buffer, while uTP strives (and manages) to limit the queue size.

These behaviors translate into different completion times statistics and, especially, completion times appear to benefit from a mixture of TCP and uTP traffic. We point out that, in the mixed case where BitTorrent peers are able to speak both protocols (*bt.transp_disposition*=31), the following happens: two connections, a TCP and an uTP

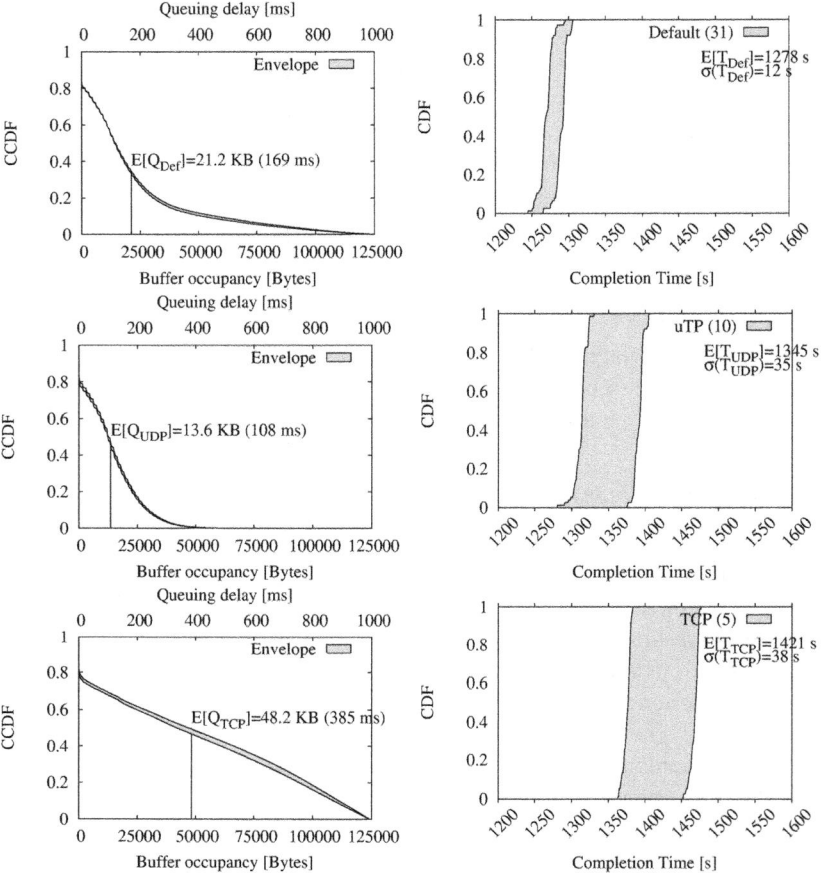

Fig. 2. Buffer occupancy CCDF (left) and Completion time CDF (right) for homogeneous swarms: default settings (*bt.transp_disposition*=31, top), uTP only swarm (*bt.transp_disposition*=10, center) and TCP only swarms (*bt.transp_disposition*=5, bottom). The vertical line in the Buffer occupancy plot represent the average of the queue length (in KB and ms).

ones are attempted, and in case uTP is successfully opened, it is preferred over the TCP one. This translates into a traffic mixture where about 80% of the data traffic happens to be carried over uTP.

Notice that the queue size alone cannot explain the difference in the completion time statistics (as otherwise, completion time in all-uTP swarms will be the lowest). Hence, we conjecture this result to be the combination of two effects –on the control and data plane– that are assisted by the use of uTP and TCP respectively. First, a longer queue size due to TCP can negatively influence the completion time, by hindering a timely dissemination of *control information* (e.g., chunk interest). The longer the time needed to signal out interests, the longer the time prior to start their download, and their subsequent upload to other peers (which harms all-TCP completion time).

Notice indeed that as in the all-TCP case the one way queuing delay may reach up to 400 ms on average, this entails that RTT for signaling exchanges may be on the order of a second, that can possibly slow down significantly the chunk spreading dynamics. Consider then that BitTorrent is using pipeling to avoid a slowdown of the transfer due to the propagation delay of requests for new chunks. From our experiments, it appears that the pipelining used by BitTorrent is not large enough to deal with delays that might be encountered with xDSL connections.

However, as previously said, the completion times statistics are not fully explained in terms of the queuing delay, as otherwise all-uTP swarms should be the winner. Yet, while uTP limits the queue size and avoids to interfere with a timely dissemination of control messages, uTP is also by design less aggressive than TCP. It follows that TCP may be more efficient for rapidly sharing *chunks in the data plane*. This can in turn harm the all-uTP completion time, that is slightly larger with respect to the default settings *bt.transp_disposition*=31.

Interestingly, our previous simulation study [18] shown that a combination of TCP and uTP can increase the efficiency on the case of two flows sharing a bottleneck link. Shortly, this happens because the low-priority protocol is still able to exploit the capacity unused by TCP (as its rate increases when queuing is low), without at the same time increasing the average system queuing delay (as its rate slow down when TCP traffic increases). The experimental results of this work further confirm that a combination of TCP and uTP can be beneficial to the completion time of the whole swarm as well. Moreover, although the exact shape of the completion time CDFs differ across experiments (due to the stochastic nature of BitTorrent chunk scheduling and peer selection decisions), the results are consistent across all iterations.

Unfortunately, latest versions of uTorrent do not allow to export chunk level logs, which could bring further information as the trading dynamics between peers, that remains an interesting direction for future work.

4.3   Heterogeneous *bt.transp_disposition* Settings

Having seen that a mixture of TCP and uTP protocols can be beneficial to the completion time, we further investigate different shares of TCP (*bt.transp_disposition*=13) vs uTP peers (*bt.transp_disposition*=14), i.e., peers that prefer one of the two protocols for active connection open, but that can otherwise accept any incoming connections.

We consider three peer-wise shares, namely 25/75, 50/50, and 75/25 (in the X/Y notation, $X\%$ represents the percentage of peers preferring TCP on their uplink, i.e., *bt.transp_disposition*=13). These shares represent three different popularity cases of uTP, that can be either the default in only few implementations of the BitTorrent applications (25/75), or compete equally (50/50) or even be dominant (75/25). We believe these shares to represent illustrative points, covering all relevant scenarios of the possible population repartition.

The plots in Fig. 3 additionally report (i) the average queuing delay, for all the swarm as well as for different peer classes, (ii) the peer- and byte-wise traffic shares, and (iii) the average system completion time, as well as the average completion time for peers of different classes.

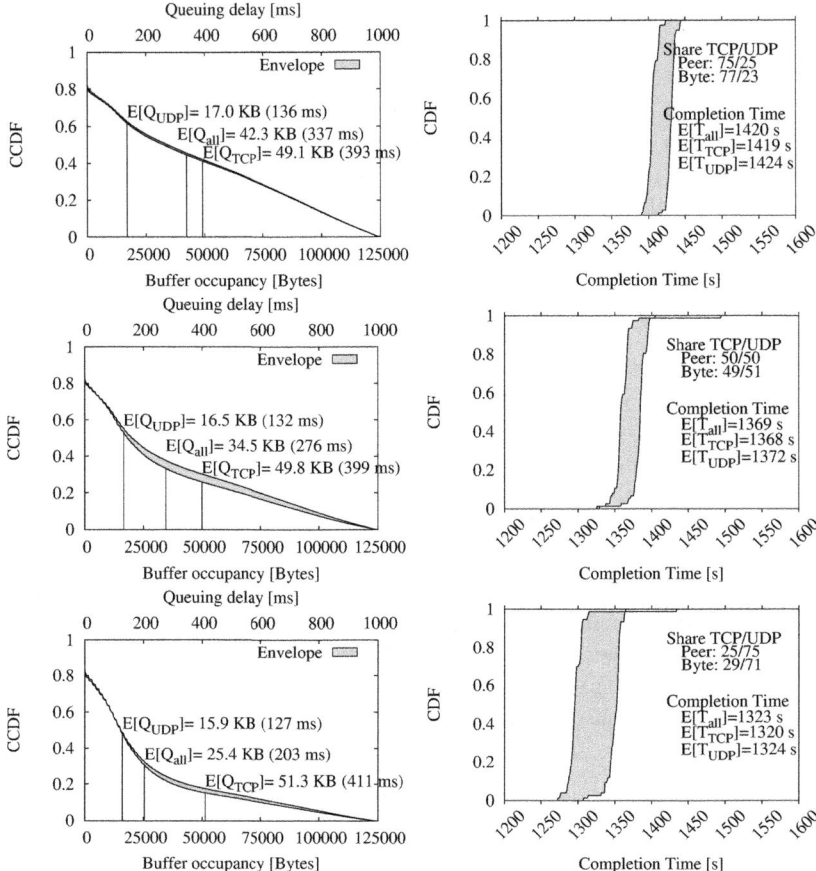

Fig. 3. Buffer occupancy CCDF (left) and Completion time CDF (right) for heterogeneous swarms: prevalence of TCP peers (75/25, top), fair population share (50/50 middle), and prevalence of uTP peers (25/75, bottom)

As for (i), the average queuing delay statistics are as expected, with an increase of the queuing delay of uTP peers due to bursty acknowledgements in reply to TCP traffic due to TCP peers in the reverse path. As for (ii), the byte-wise share closely follow the peer-wise share. Finally, let us focus on (iii) the completion time statistics. Interestingly, as Fig. 3 shows, while a small amount of TCP traffic is beneficial in reducing the overall swarm completion time (bottom row), a large TCP amount can instead slow down the torrent download for the whole system (top row).

Further, notice that completion times are practically the same for uTP and TCP peers (with a slight advantage for the latter). Hence, differently from our previous simulation work [9], we do not observe an unfairness of completion time between different peer classes within an heterogeneous swarm. This is due to the fact that [9] considered a simpler model for *bt.transp_disposition*, that neither (i) accounted for TCP peers using an already opened uTP connection in the reverse side nor (ii) for the hardcoded uTP preference.

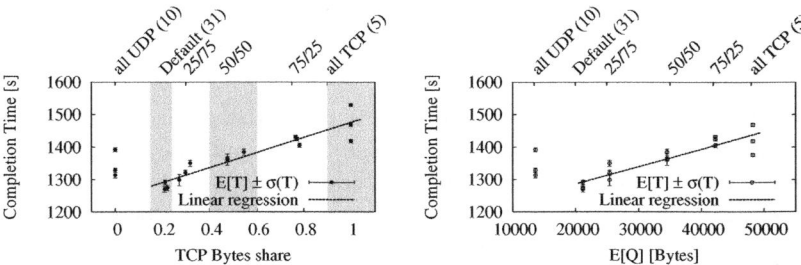

Fig. 4. Completion time as a function of the TCP vs uTP byte-wise share (left) and as a function of the average buffer occupancy (right)

4.4   uTP vs TCP in a Nutshell

Fig. 4 present a summary of our results considered so far. T is the completion time (mean and standard deviation) for different iterations with both homogeneous and heterogeneous swarm populations. The T metric is reported as a function of the byte-wise TCP traffic share (left plot) and of the average buffer occupancy (right plot).

Both plots also report, for TCP traffic shares different from zero (non-0 TCP) only, a linear regression of the completion time. Notice that the linear model provides a nice fit to forecast the completion time performance in presence of different TCP vs uTP mixtures.

Furthermore, as observed in [9] by means of simulation, Fig. 4 confirms that the *completion time increases for increasing buffer occupancy,* which in turn generally increases with the amount of TCP traffic exchanged.

As previously argued, this is due to a slow-down of BitTorrent signaling traffic, while the completion time increase of all-uTP swarms is instead likely due to the low-priority of uTP in the data plane. Hence, we also remark a non-monotonous behavior for the completion time, that decreases for decreasing percentages of TCP traffic, and then increases again for all-uTP swarms. As different dynamics takes place, hence the linear dependence only applies in case of uTP and TCP traffic mixtures (i.e., non-0 TCP traffic share).

Finally, notice that the default BitTorrent settings consistently yield to the shortest download time we observed in the experiments, which confirms the soundness of the *bt.transp_disposition* design decision and settings.

5   Conclusions

This work assess the impact of uTP (the new BitTorrent congestion control algorithm for data exchange) on the torrent completion time (the main user QoE metrics) by means of an experimental campaign carried on in a fairly large scale controlled testbed.

Our results show that, in flash crowd scenarios, users will generally benefit of a mixture of TCP and uTP traffic, both in homogeneous and heterogeneous swarms. Interestingly indeed, results with mixed TCP and uTP traffic show consistently shorter download time with respect to the case of homogeneous swarms using either an all-TCP or an all-uTP congestion control. Especially, our results confirm the soundness of default BitTorrent settings, which use both TCP and uTP protocols and lead to the shortest completion time in our experiments.

This results is the combination of two effects, on the control and data plane, that are assisted by the use of uTP and TCP respectively. By keeping the queue size low, uTP yields to a timely dissemination of signaling information, that would otherwise incur in higher delays due to longer queues building up with TCP. At the same time, by its more aggressive behavior, TCP yields to higher efficiency in the data plane, that results in more timely dissemination of chunk content.

This work leaves a number of interesting points open, that we aim at addressing in the future. First, we would like to investigate whether it would be possible to extend the swarm size by running multiple peers per machine, without however incurring in a bias due to the mutual interaction of the traffic injected. Second, we would like to study the impact of heterogeneous target values in the swarm (i.e., to see whether fairness issues can possibly arise) and refine our experimental setup (e.g., including heterogeneous capacities, peer churn, etc.). Third, on a longer timescale, we aim at developing a passive BitTorrent inspector, capable of parsing traffic to produce chunk-level logs, that would greatly enhance our analysis capabilities.

Acknowledgement. We thank Arvid Norberg for the fruitful discussions. Experiments presented in this paper were carried out using the Grid'5000 experimental testbed, being developed under the INRIA ALADDIN development action with support from CNRS, RENATER and several Universities as well as other funding bodies [10]. Project funding for our future work on the topic would be more than welcome.

References

1. Alcock, S., Nelson, R.: Application flow control in youtube video streams. ACM SIGCOMM Computer Communication Review 41(2), 24–30 (2011)
2. Bonfiglio, D., Mellia, M., Meo, M., Rossi, D.: Detailed analysis of skype traffic. IEEE Transaction on Multimedia 11(1), 117–127 (2009)
3. Shalunov, S.e.a.: Low Extra Delay Background Transport (LEDBAT). IETF Draft (October 2010)
4. Norberg, A.: BitTorrent Enhancement Proposals on μTorrent transport protocol (2009), http://www.bittorrent.org/beps/bep_0029.html
5. IETF: LEDBAT Working Group Charter, http://datatracker.ietf.org/wg/ledbat/charter/
6. Schneider, J., Wagner, J., Winter, R., Kolbe, H.: Out of my Way – Evaluating Low Extra Delay Background Transport in an ADSL Acciess Network. In: 22nd International Teletraffic Congress (ITC22), Amsterdam, The Netherlands (September 2010)
7. Rossi, D., Testa, C., Valenti, S.: Yes, We LEDBAT: Playing with the New BitTorrent Congestion Control Algorithm. In: Krishnamurthy, A., Plattner, B. (eds.) PAM 2010. LNCS, vol. 6032, pp. 31–40. Springer, Heidelberg (2010)
8. Morris, S.: μTorrent release 1.9 alpha 13485 (December 2008), http://forum.utorrent.com/viewtopic.php?pid=379206#p379206

9. Testa, C., Rossi, D.: The impact of utp on bittorrent completion time. In: 11th IEEE International Conference on Peer-to-Peer Computing (P2P), Kyoto, Japan (2011)
10. https://www.grid5000.fr
11. Venkataramani, A., Kokku, R., Dahlin, M.: TCP Nice: A mechanism for background transfers. In: 8th USENIX Symposium on Operating Systems Design and Implementation (OSDI 2002), Boston, MA (December 2002)
12. Kuzmanovic, A., Knightly, E.: TCP-LP: low-priority service via end-point congestion control. IEEE/ACM Transactions on Networking (TON) 14(4), 752 (2006)
13. Liu, S., Vojnovic, M., Gunawardena, D.: 4cp: Competitive and considerate congestion control protocol. In: ACM SIGCOMM, Pisa, Italy (September 2006)
14. Qiu, D., Srikant, R.: Modeling and performance analysis of BitTorrent-like peer-to-peer networks. ACM SIGCOMM Comp. Comm. Rev. 34(4), 367–378 (2004)
15. Rao, A., Legout, A., Dabbous, W.: Can realistic bittorrent experiments be performed on clusters? In: 10th IEEE International Conference on Peer-to-Peer Computing (P2P), pp. 1–10 (August 2010)
16. Legout, A., Liogkas, N., Kohler, E., Zhang, L.: Clustering and sharing incentives in bittorrent systems. In: Proc. of ACM SIGMETRICS 2007, San Diego, CA, USA (June 2007)
17. Cohen, B., Norberg, A.: Correcting for clock drift in uTP and LEDBAT. Invited talk at 9th USENIX International Workshop on Peer-to-Peer Systems (IPTPS 2010), San Jose, CA (April 2010)
18. Rossi, D., Testa, C., Valenti, S., Muscariello, L.: LEDBAT: the new BitTorrent congestion control protocol. In: 19th IEEE International Conference on Computer Communications and Networks (ICCCN 2010), Zurich, Switzerland (August 2010)
19. Carofiglio, G., Muscariello, L., Rossi, D., Valenti, S.: The quest for LEDBAT fairness. In: IEEE Global Communication (GLOBECOM 2010), Miami, FL (December 2010)
20. Carofiglio, G., Muscariello, L., Rossi, D., Testa, C.: A hands-on Assessment of Transport Protocols with Lower than Best Effort Priority. In: 35th IEEE Local Computer Network (LCN 2010), Denver, CO (October 2010)
21. Bhooma, P.: Measurements using the LEDBAT implementation for MAX OS X, http://www.ietf.org/mail-archive/web/ledbat/current/msg00532.html
22. Abu, A., Gordon, S.: A Dynamic Algorithm for Stabilising LEDBAT Congestion Window. In: 2nd IEEE International Conference on Computer and Network Technology (ICCNT 2010), Bangkok, Thailand (April 2010)
23. http://perso.telecom-paristech.fr/testa/pmwiki/pmwiki.php?n=Main.BTtestbed
24. Mogul, J.C.: Tcp offload is a dumb idea whose time has come. In: 9th Conference on Hot Topics in Operating Systems, vol. 9, p. 5. USENIX Association, Berkeley (2003)
25. Akella, A., Seshan, S., Shaikh, A.: An empirical evaluation of wide-area internet bottlenecks. In: Proc. of the 3rd ACM SIGCOMM Conference on Internet Measurement (IMC 2003), Miami, FL, USA (October 2003)

Steps towards the Extraction of Vehicular Mobility Patterns from 3G Signaling Data

Pierdomenico Fiadino[1], Danilo Valerio[1], Fabio Ricciato[1,3],
and Karin Anna Hummel[2]

[1] Forschungszentrum Telekommunikation Wien (FTW), Austria
[2] ETH Zurich, Switzerland
[3] University of Salento, Italy

Abstract. The signaling traffic of a cellular network is rich of information related to the movement of devices across cell boundaries. Thus, passive monitoring of anonymized signaling traffic enables the observation of the devices' mobility patterns. This approach is intrinsically more powerful and accurate than previous studies based exclusively on Call Data Records as significantly more devices can be included for investigation, but it is also more challenging to implement due to a number of artifacts implicitly present in the network signaling. In this study we tackle the problem of estimating vehicular trajectories from 3G signaling traffic with particular focus on crucial elements of the data processing chain. The work is based on a sample set of anonymous traces from a large operational 3G network, including both the circuit-switched and packet-switched domains. We first investigate algorithms and procedures for preprocessing the raw dataset to make it suitable for mobility studies. Second, we present a preliminary analysis and characterization of the mobility signaling traffic. Finally, we present an algorithm for exploiting the refined data for road traffic monitoring, i.e., route detection. The work shows the potential of leveraging the 3G cellular network as a complementary "sensor" to existing solutions for road traffic monitoring.

1 Introduction

Road congestions cause safety dangers, unwanted CO_2 emissions, and major economic losses. Informing the drivers about real-time traffic status helps to reduce individual travel times, to optimize traffic flows, and to make a more efficient use of the road infrastructure. The systems currently used for gathering information about road traffic conditions are mainly based on road sensors, which would be very expensive to deploy on a large scale. Therefore, only some critical road sectors and junctions are covered by such monitoring infrastructure.

Cellular networks can help to overcome this problem: millions of mobile devices held by car drivers and passengers can be used opportunistically as road traffic probes, without facing the costs of deploying any new infrastructure. The main goal of this work is to develop a system that extracts road mobility data from the signaling traffic in a cellular network and uses the obtained information both for real-time applications and historical data analysis. Real-time applications include, e.g., inference of current road traffic intensity, detection of

A. Pescapè, L. Salgarelli, and X. Dimitropoulos (Eds.): TMA 2012, LNCS 7189, pp. 66–80, 2012.
© Springer-Verlag Berlin Heidelberg 2012

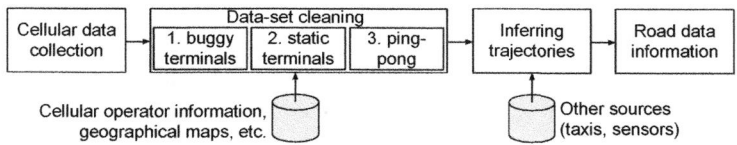

Fig. 1. Vehicular traffic monitoring approach including cellular data

congestions, and estimation of expected travel times. Historical data analysis can provide support for tasks like road infrastructure planning, urban planning, optimization of public transport, etc.

The development of such a system must face several challenges. First, cellular networks produce an enormous amount of signaling data. Thus, efficient methods are needed to filter out data that are not relevant for the target task, e.g., from static or off-road terminals. Second, the mobility *perceived* by the network does not always map directly to the *actual* geographical mobility of the devices, due to the dynamics of Mobility Management (MM) protocols and radio propagation effects (e.g., fading). Third, cellular protocols are designed to minimize the signaling traffic through the network. Devices with no data or voice activity are tracked at a coarser spatial accuracy than active ones, i.e., at the level of location/routing area instead of cell. The system must be able to cope with this aspect, e.g., by reconstructing the state of the device by finite state machine.

Taking into account these challenges, we present a flexible traffic monitoring infrastructure on top of the cellular network that (*a*) collects signaling data from the network probes, (*b*) pre-processes the data and filters out irrelevant and/or spurious information, and (*c*) infers mobility trajectories across roads in order to pave the way towards the extraction of mobility patterns. In this paper, we describe the steps necessary to extract mobility patterns and include a desription of the artifacts emerged during the analysis of real-world cellular data.

The remainder of this paper is organized as follows: In Section 2, we describe the monitoring framework and present the main components of the cellular network architecture. In Section 3, we explore important examples of pre-processing filters that are required for reliably extracting mobility patterns. In Section 4, we present a first analysis and characterization of the dataset, and in Section 5 we introduce an algorithm for trajectory estimation. In Section 6, we relate our approach to other works. Finally, in Section 7 we draw the main conclusions.

2   Monitoring Framework

The analysis in this work was conducted on a dataset of anonymized signaling traffic captured during an entire working day in an Austrian cellular network. Figure 2 shows the main components of the monitoring system along with an overview of the cellular network architecture.

The cellular infrastructure is composed of a Core Network (CN) and a Radio Access Network (RAN). The CN is divided in two distinct domains: Circuit-Switched (CS) and Packet-Switched (PS). Mobile terminals can "attach" to the CS for voice call services, to the PS for packet data transfer, or to both. They

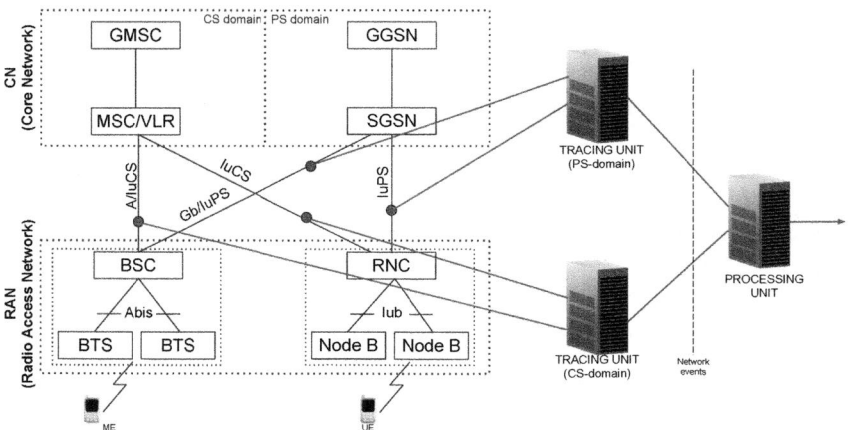

Fig. 2. Monitoring framework integrated into a cellular network architecture

can do so from 2G (GSM/EDGE) or 3G (UMTS/HSPA) radio bearers. Radio communication occurs between a mobile terminal and a base station serving a cell, which is the smallest spatial entity in the cellular network. Cells are grouped into larger logical entities: Routing Areas (RAs) and Location Areas (LAs) for the PS and CS domain respectively. In order to remain reachable, the terminals always inform the CN whenever they change LA and/or RA.

Our monitoring system collects signaling from the links between the cellular RAN and CN as indicated in Fig. 2, specifically on IuPS, IuCS, Gb and A interfaces. The amount and accuracy of mobility information that can be gathered from such interfaces varies with the terminal state: in general, the location of the terminals with an ongoing voice or data connection (active terminals) is known at the cell level, while the location of idle terminals is known only at LA/RA level. The network probes are connected to a couple of tracing units that interpret the signaling traffic and generate event-based tickets that are then forwarded to a processing unit. These tickets do not contain any identification of the devices nor payload of user traffic. In order to correlate and aggregate trajectories, each device is assigned an anonymous fingerprint generated by a one-way hash function of the corresponding International Mobile Subscriber Identity (IMSI) number. To further preserve user privacy, the hash key is regenerated daily with different seeds. For additional details about the monitoring system and the relevant MM protocols refer to [1,2]. A detailed overview of the cellular network technology can be found in [3].

3   Mobility and Dataset Filtering

The ultimate goal of this work is to estimate vehicular trajectories based on (i) the passively monitored signaling traffic and (ii) the (known) geographical position and antenna orientation of the base stations. Hereby, defining a trajectory is not trivial: base station deployment parameters — e.g., directions and ranges of the antennas — and MM protocol dynamics must be taken into account to map cell sequences to a geographical trajectory. The raw dataset must undergo some preliminary filtering steps, namely:

(*1*) filtering traffic from buggy terminals and data inconsistencies;
(*2*) identifying and filtering traffic from static terminals;
(*3*) filtering network mobility artifacts.

Given the complexity of the monitored network, one cannot assume that traces collected in real-world networks exhibit only expected signaling patterns: anomalies caused by buggy terminals, data holes, and other effects due to the limitations of the monitoring system cannot be avoided. Buggy behaviors can be caused by crashes or freezes in the operating system or in the baseband processor firmware of the terminals, causing the device to generate repeatedly the same message to the network. Such spurious traffic pollutes the dataset with sequences that interfere with the data processing algorithms. This traffic can be easily filtered out by simple thresholding, i.e., filtering all devices generating more than i events in a time interval t (e.g., per minute), as done in [4].

Cellular networks serve a large share of static terminals that do not change location, e.g., home computer with 3G modem. Considering the enormous amount of data to process, it is convenient to eliminate such devices from the dataset so as to relax the requirements in terms of processing resources, without any impact on the final output quality. Again, a simple thresholding approach is sufficient, filtering out devices that visit less than c cells in a time interval t.

The mobility *perceived* by the network signaling does not always map to the *real* geographical mobility of the devices. One of the main effects that must be considered is the presence of so-called "ping-pong" patterns, i.e., the repeated handover of a device between two or more cells caused by fluctuations of the signal strength level in time and/or space due to fading. Terminals exposed to the ping-pong effect are perceived by the system as highly mobile, but actually they can be geographically static. It is convenient to eliminate such cases to prevent distortions in the route classification algorithms presented later in Section 5.

For this preliminary study we used a simple algorithm for the automatic detection and filtering of ping-pong events. Given two cells A and B, we define a *ping-pong generated hop* when a device moves from cell A to B, and back, within a time interval t. The script keeps track of the last two cells crossed by each terminal. Assuming that a device generates a signaling message in the n-th cell, the system knows the $n-1$ and $n-2$ previous cell-ids. If cell n differs from cell $n-1$, the script checks whether cell n is equal to cell $n-2$. If so the timestamps of these two cells are checked: if $timestamp_n - timestamp_{n-2} < t$ the hop is considered to be part of a ping-pong sequence and therefore it is filtered. The algorithm can be easily extended to detect ping-pongs among more than two cells. Some care must be taken in choosing the right time window t. Too large values might cause false positives, hence discarding of useful data of local mobility. For this study, where the focus is on highway scenarios, we used a conservative setting of $t = 1$. In fact, due to the high speed of highway users,

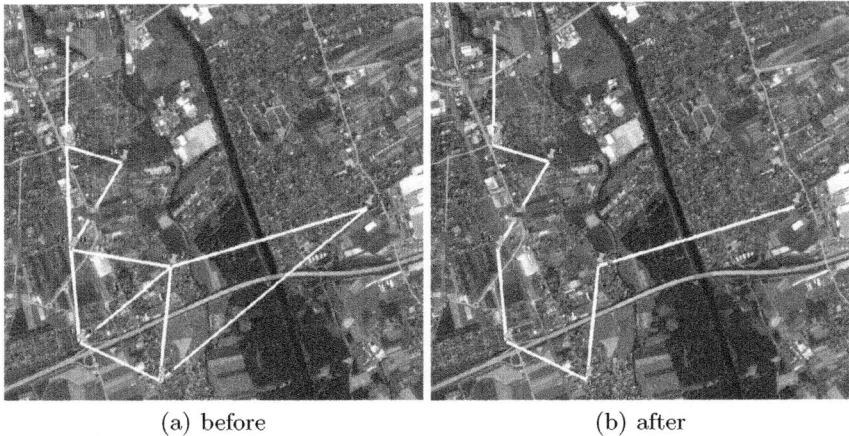

(a) before   (b) after

Fig. 3. Visualization of the ping-pong effect before (left) and after filtering (right)

fading fluctuations seen by such devices are rapid — also fading at large-scale in space maps to fluctuations at small-scale in time — hence most cell transitions due to ping-pong have short durations. Instead, identifying the optimal setting of t for urban areas is more challenging — we leave this point for further study.

We tested the algorithm experimentally on the road. Figure 3 shows an example of how one of our test-drives is perceived by the network. The test car was equipped with a 3G handset and a GPS device while driving on a highway exit ramp. The picture on the left depicts the cell sequence seen in the raw dataset. The sequence contains 17 hops occurred in 8 different cells. On the right side the same trajectory is depicted after filtering the ping-pong effect. About 50% of the hops were caused by ping-pong effects, which could be filtered by our algorithm leading to a visibly smoother path.

4   Signaling Traffic Characterization

Given the complexity of the analysis task, it is convenient to familiarize with some basics aspects of the dataset at hand before delving into the algorithm details, e.g., looking at simple distributions of traffic across cells and/or terminals and exploring the principal different classes of terminal behaviour. In this section, we present some basic aspects of the MM signaling present in our sample dataset from a real-world cellular network in Austria. Where applicable, we show the effects of applying the filtering steps discussed in the previous section.

4.1   Time-Charts of Signaling Events

Figure 4 shows the time series of the number of signaling events during an entire day, aggregated per minute[1]. The chart on the left shows the total amount of

---

[1] In this and the following plots the values in the y-axis are normalized to prevent disclosure of data considered business-sensitive by the network operator.

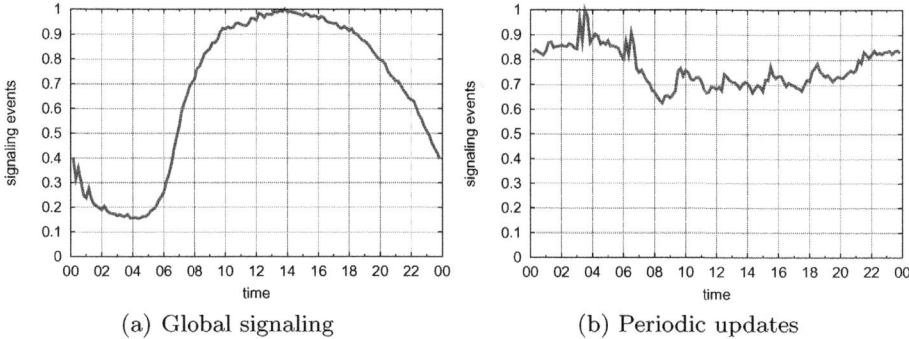

(a) Global signaling   (b) Periodic updates

Fig. 4. Signaling traffic observed in one day

signaling events like cell handovers, LA updates, etc. The typical time-of-day pattern associated to the daily cycle of human behaviour is clearly visible.

On the right side we plot the time series of the *periodic LA updates*, i.e., a sort of "keep-alive" message that is sent by terminals that do not move nor engage in any active voice/data call for a certain time. Therefore, such traffic is anti-correlated with user mobility — the less they move, the more likely they produce periodic LA updates — and in fact the trend is somewhat opposite to the previous graph.

4.2   Number of Crossed Cells

The solid line in Figure 5 shows the number of unique cells visited in one day by each terminal, considering all signaling messages in the dataset. The plot shows that more than 15% of terminals appear only in a single cell, while approximately 50% generate signaling in less than five cells. This confirms the opportunity of introducing a preprocessing stage to remove static users from the raw dataset.

The dashed line emulates what would be seen by CDR: it reports the number of different cells where the user started or terminated a voice/data connection. More than 25% of the terminals generate/receive calls in a single cell during a

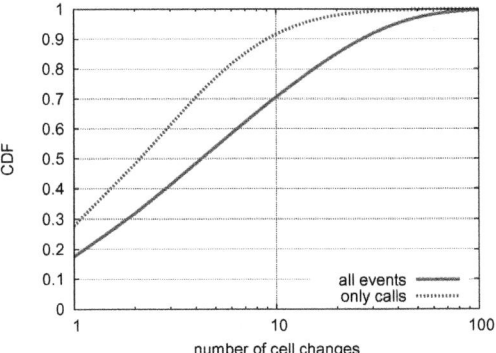

Fig. 5. Number of daily crossed cells

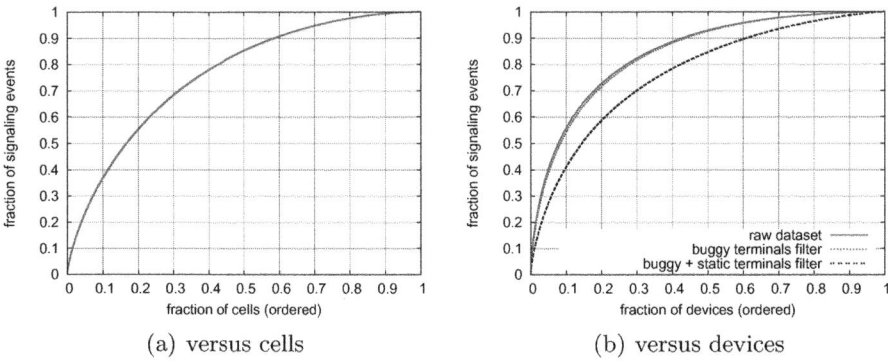

Fig. 6. Cumulative distribution of signaling traffic

whole day. The remarkable gap between the two curves shows that relying on CDRs produces a biased perception of the mobility of the population, leading to underestimate the actual level of user mobility. In other words, we conjecture that the *quantitative metrics* reported in CDR-based studies like, e.g., [5,6], are not representative of the actual mobility process. A more accurate comparison between the mobility measurements obtained by CDR and CN/RAN signaling is part of our ongoing work.

4.3 Signaling Distribution across Cells and Subscribers

Figure 6(a) shows the distribution of the signaling traffic across cells. The plot reveals the existence of a small set of high-activity cells: 30% of the base stations are involved in the exchange of 70% of the signaling traffic. Analogously, Figure 6(b) depicts the distribution across terminals: considering the raw dataset, 20% of the devices attached to the network generate 75% of the total amount of signaling traffic. These types of distributions with large disparity are often found in cellular networks [7,8,9,10,11,12]. The question arises whether this distribution is representative for the signaling traffic even after filtering out buggy terminals and static users. The plot shows that the removal of buggy terminal does not have a remarkable effect on the distribution, thus the curves are almost overlapping. In fact, while these terminals have deleterious effects in the processing algorithms, they are too sparse to influence the total distribution. The same cannot be said for static users. When our filtering algorithm is applied (dotted line) the curve changes considerably, showing a more uniform distribution.

4.4 Idle vs. Active Devices

The previous section highlighted the existence of a large set of static mobile devices and a small set of highly mobile ones. Figure 7 shows the distribution of *cell update* signaling messages among devices. Such messages are produced only by user in active state, engaged in data and/or voice connection. It can be seen that there is a small class of terminals that generate a very large number of cell updates. These terminals are always active and provide the network with fine-grained information about their mobility.

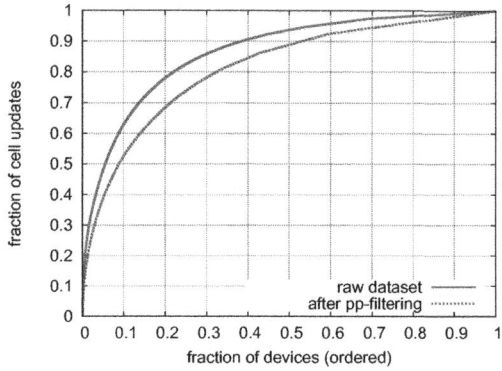

Fig. 7. Distribution of *cell updates* among devices

An efficient system for road traffic flow analysis must consider mobility events from both idle and active users. At any time the vast majority of the terminals are in idle state. These provide a more complete view of the mobility flows than the fewer active terminals, but at a coarser spatial granularity (LA or RA level). Conversely, active terminals provide detailed information about their movements at cell level and can be used as single, sparse, and more accurate road traffic probes in addition to passive terminals. In some sense the two classes play complementary roles in terms of sample coverage vs. spatial accuracy. Thus, in our approach we exploit data from both active and idle users to develop a system that dynamically reveals traffic jams, road anomalies, and travel times more accurately than alternative approaches based only on signaling traffic from active users such as [13,14].

5   Route Detection

Route detection is a prerequisite for several road traffic studies based on cellular network signaling. The goal of route detection is to map handset transitions (handovers) among network entities to driver trajectories in the road network.

In this section, we propose a novel algorithm that is based on the *vector space classification* and takes inspiration from the field of Web information retrieval and relevance ranking. The key idea is the following: the sequence of base stations covering a path (road) and the sequence of cells crossed by terminals can be represented as vectors in a vector space. In this context, the problem of deciding whether a cellular user is traveling along a road is solved with the computation of the *similarity* of the respective user and road vectors: the higher the similarity, the higher the probability that the cellular user traveled along the corresponding road. The procedure for performing route detection is composed mainly by two steps: *vector generation* and *similarity computation*.

5.1   Vectors Generation

We define a *road descriptor* as a static sequence of cells that cover the road in a specific direction of interest. In other words, the road descriptor represents

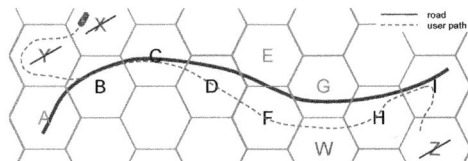

Fig. 8. Example of a user's cell sequence

the sequence of cells which a device is most likely to communicate with, when traveling the road in one direction. Road descriptors can be built by using information available in radio network planning departments of mobile operators or by simple test-drives aimed at building cell-handovers logs. Note that the sequence of cells differs depending whether the analysis focuses on idle or active devices, i.e., in case of idle devices it includes only cells at the LA borders.

We define a *user path descriptor* as the sequence of cells traversed by a terminal. This sequence starts being computed as soon as the device attaches to a cell belonging to a road descriptor. Figure 8 shows an example: the road descriptor is (A,B,C,D,E,F,G,H,I), the terminal covers cells (X,Y,B,C,D,F,W,H,I,Z) and the resulting user path descriptor is (B,C,D,F,W,H,I).

The two descriptors are used to generate road and user vectors in the *n-dimensional* space, where n is the total number of unique cells in road and/or user sequence. Both vectors are binary and of the same dimension. The generation of user path descriptors and the procedure for building road and user vectors are formally described by pseudo-code in Algorithm 1 and Algorithm 2. Figure 9 shows an example of road and user vectors, where the road sequence is composed of 5 cells and the user traverses 2 additional cells (i.e., $n = 7$).

5.2 Similarity Computation

The second step involves the computation of the similarity between user and road vectors. The similarity between two vectors is defined as the inverse of their geometrical distance and ranges between 0 and 1.

The similarity can be computed using several metrics. Some examples are the cosine similarity, the Jaccard index, and the Tanimoto coefficient [15]. The latter is the most appropriate for our aims, as it is designed for binary vectors and penalizes the presence of *external cells* more than other similarity metrics. In other words, it mitigates better the problem of ambiguous user placement in case of nearly parallel roads. Tanimoto similarity is defined as

$$T(A, B) = \frac{A \cdot B}{\|A\|^2 + \|B\|^2 - A \cdot B}$$

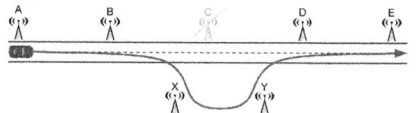

	A	B	C	X	Y	D	E
Road vector	1	1	1	0	0	1	1
User vector	1	1	0	1	1	1	1

Fig. 9. Example of vectors generation

Algorithm 1. UserSequenceExtraction (*user_sequence u, road_sequence r*)

1: $k \leftarrow 0$
2: $start\_flag \leftarrow 0$
3: **for** $i = 1 \dots |u|$ **do**
4:     **if** $u[i] \in r$ **then**
5:         $start\_flag \leftarrow 1$
6:         $k \leftarrow$ position of $u[i]$ in r
7:         **if** $j > k$ **then**
8:             $k \leftarrow j$
9:             append aux_vector to $result\_sequence_{(u,r)}$
10:             append $u[i]$ to $result\_sequence_{(u,r)}$
11:             empty $aux\_sequence$
12:         **end if**
13:     **else**
14:         **if** $start\_flag = 1$ **then**
15:             append $u[i]$ to $aux\_sequence$
16:         **end if**
17:     **end if**
18: **end for**
19: **return** $result\_sequence_{(u,r)}$

---

where A and B are two binary vectors and $A \cdot B = \sum_i (A_i \wedge B_i)$, $\|A\|^2 = \sum_i (A_i)$. We computed the Tanimoto similarity with the following equation:

$$similarity = \frac{(\# \; internal \; cells)}{(length \; of \; road \; descr) + (length \; of \; user \; seq) - (\# \; internal \; cells)}$$

thus, without the vector generation procedure but directly using road and user sequences. This allows us to release processing power for other tasks. Note that all previous considerations about vectorial route detection remain valid even when this simplified formula is applied.

5.3   Route Classification

Once all similarities have been computed, each user is assigned to the road whose vector presents higher similarity. This implies that there exists a road descriptor for each road in the region of interest. In reality, extracting road descriptors is a tedious procedure that requires a number of test-drives and detailed knowledge of the operator radio planning. Thus, it is not always possible to derive road descriptors for the entire road network. To overcome this problem we propose a threshold-based metric, i.e., a road sector is considered to be traveled by a terminal when the similarity between the user and the road vector is higher than a specific threshold. In the experiments we have performed so far, we manually tuned this parameter and analyzed the type and number of output trajectories. Validating such results is difficult due to the lack of ground truth to compare with. Note that according to the vector generation procedure described above, the similarity is computed for every pair of user-road vectors that share at least one common cell in the descriptors. Figure 10 shows an illustrative example of

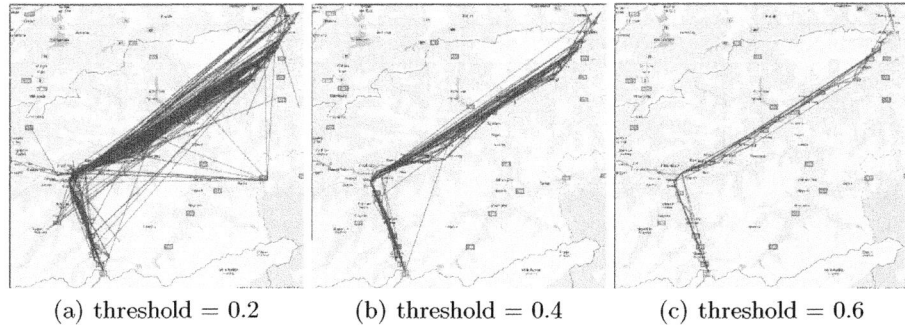

(a) threshold = 0.2   (b) threshold = 0.4   (c) threshold = 0.6

Fig. 10. Comparison of different similarity thresholds

Algorithm 2. VectorsGenerator (*filtered_user_sequence u, road_sequence r*)

1: $i \leftarrow 0$
2: $j \leftarrow 0$
3: **while** $i < |r|$ **do**
4:    **if** $r[i] \neq u[j]$ **then**
5:       **while** $u[j] \notin r$ **do**
6:          append 0 to *road_vector*
7:          append 1 to *user_vector*
8:          $j + +$
9:       **end while**
10:      **if** $u[j] = r[i]$ **then**
11:         append 1 to *user_vector*
12:      **else**
13:         append 0 to *user_vector*
14:      **end if**
15:   **else**
16:      append 1 to *user_vector*
17:      $j + +$
18:   **end if**
19:   append 1 to *road_vector*
20:   $i + +$
21: **end while**
22: **return** *user_vector, road_vector*

---

the effect of different threshold values. We compute the similarity of every mobile users' vector with a single sample road vector representing part of an Austrian motorway. It can be noted that the lower the threshold, the higher the amount of vehicles associated with the road segment. A too low value increases the probability of including users who didn't actually driven that road. A too high threshold reduces considerably the number of output trajectories, but potentially excludes some of the target users. The optimal value of the threshold depends on several factors, including the accuracy of the road vector, the radio network design, the existence of overlapping cells and umbrella cells, the length of the considered road segment, etc. Automatic tuning of this parameter is part of our ongoing work.

6   Related Work

The use of cellular networks for extracting vehicular mobility data has been the focus of several commercial products in the last years, aiming at selling cheaper and more scalable ways to infer and redistribute road traffic information [16,17,18]. Unfortunately, given their commercial nature, no public information is available describing algorithms and procedure used by these products, preventing the scientific community from analyzing and evaluating such systems.

A few reserch works and projects have been devoted to the topic and only a subset of them utilizes real-world data from operational mobile networks. In general, one can distinguish between active, passive, and application-based mobility monitoring. The most common systems rely on the application-based approach, i.e., smartphone applications that report GPS and cellular information to a central server. In [19,20] the authors propose an application-based system for road traffic estimation using a small set of sample users. Sricharan et al. [21] use the same approach for the characterization of users' mobility and clustering in homogeneous groups. Analyzing a large amount of user reported GPS information is the aim of other projects such as the Mobile Millennium Project [22]. GPS data is particularly suitable for studying micro-mobility characteristics, including velocity, directional changes, and start time of trips, as shown in our previous work [23], but data sets are usually smaller than those from cellular networks.

Most of previous studies based on passive data from the cellular network make use of (anonymized) Call Data Records (CDRs) [5,6,7,8,9]. CDR are summary tickets produced for billing purposes at every voice call, data connection or SMS envoy. With CDR the user location can be observed only at the time of initiating or terminating a call, data connection or SMS. In general, this approach gives a limited and somewhat biased view of user mobility, as his trajectory is sampled at instants that are dependent on (conditioned to) his calling activity, which in turn might depend on factors like time-of-day, position etc. The limitation of the CDR approach is particularly serious when considering road users, that in general are less likely to engage in calls during the trip.

Trestian *et al.* [13,14] adopted a passive monitoring approach that is somehow intermediate between the CDR method and the one we adopt in this work. By monitoring the CN links in the PS domain they are able to observe the user location only during data connections, i.e., for active terminals. Instead, by monitoring the links between the CN and the RAN, we collect mobility information also for terminals in idle state. Furthermore, our dataset covers both the PS and CS domains, and both 2G (GSM/EDGE) and 3G (UMTS/HSPA) radio access.

Two other projects, namely *Traffic.online* [24] and *Do-iT* [25], consider passive monitoring data from the RAN, and specifically the A-bis interface. This approach provides very accurate information but is also expensive, as it might require to tap hundreds or even thousands of A-bis links to cover a nation-wide network, depending on the particular network deployment.

Besides empirical analyses of real-world data, the literature is rich of studies based on network-level simulations. Among them, our work is closest related to the work of Gundlegard et al. [26,27] who state goals similar to ours but apply different methodologies. A detailed survey of the state-of-the-art in road traffic estimation from cellular mobility signaling can be further found in one of our previous publications [1].

7   Conclusions

In this paper, we presented a work-in-progress study on the potential of using signaling traffic from the mobile network for road mobility analysis. The work is based on a real-world dataset from an Austrian cellular network operator.

We showed the main processing steps that must be taken to extract road mobility patterns from the signaling traffic stream, highlighting the necessary pre-filtering procedures. We showed that the signaling dataset contains a low amount of highly mobile users. We also discussed the importance of considering both idle and active terminals, that provide complementary views in terms of coverage and spatial granularity. Finally, we presented a method to map "road descriptions" and "user paths", both in terms of cell sequences, by applying a distance metric for binary vectors.

As for our future work, we are currently collecting data from a set of reliable alternative sources and building an accurate ground-truth for the validation of the presented results. In particular, we are conducting several test-drives in many well-traveled Austrian motorways. In each run we log all events reported by the radio interface layer of GPS-equipped smartphones and correlate the actual mobility (i.e. GPS based) with the mobility perceived by our system in terms of traveled road sections. In addition, we are performing comparisons between the traffic intensities reported by selected road sensors of the motorway operator and the ones perceived by our system. Preliminary results are very encouraging. Finally, we are tuning the parameters of the algorithms presented throughout this paper, with particular focus on investigating and identifying the optimal threshold values for the ping-pong filter and the Tanimoto similarity. Our final aim is to design a road monitoring module that continuously analyzes cellular network signaling and infers current traffic conditions reliably and *in real-time*. We envision a system that is able to detect anomalies using particular events in the signaling traffic — e.g., a drop in the handover rate or sudden change in the number of road user across a road segment — and combine current road condition with past history in order to predict upcoming road congestions, i.e., traffic forecast.

In another parallel work we aim at comparing the different views that the three approaches to mobility estimation based on cellular network data — namely CDR [5,6], CN/PS [13,14] and our approach — can provide about the underlying human mobility patterns.

Acknowledgment. This work has been supported by the Austrian Government and by the City of Vienna within the competence center program COMET. The work of K.A. Hummel has been supported by the Commission of the European Union under the FP7 Marie Curie IEF program contract PIEF-GA-2010-276336 MOVE-R.

References

1. Valerio, D., et al.: Road traffic estimation from cellular network monitoring: a hands-on investigation. In: IEEE Personal Indoor Mobile Radio Communication Simposium, IEEE PIMRC 2009 (September 2009)
2. Valerio, D., et al.: Exploiting cellular networks for road traffic estimation: a survey and a research roadmap. In: IEEE VTC 2009 Spring (April 2009)
3. Holma, H., Toskala, A.: WCDMA for UMTS: Radio Access for Third Generation Mobile Communications. John Wiley and Sons, Ltd. (2000)
4. Ricciato, F., Pilz, R., Hasenleithner, E.: Measurement-Based Optimization of a 3G Core Network: A Case Study. In: Koucheryavy, Y., Harju, J., Iversen, V.B. (eds.) NEW2AN 2006. LNCS, vol. 4003, pp. 70–82. Springer, Heidelberg (2006)
5. González, M.C., Hidalgo, C.A., Barabási, A.L.: Understanding individual human mobility patterns. Nature 453(5) (June 2008)
6. Song, C., Qu, Z., Blumm, N., Barabási, A.L.: Limits of predictability in human mobility. Science 327(5968) (February 2010)
7. Halepovic, E., Williamson, C.: Characterizing and Modeling User Mobility in a Cellular Data Network. In: 2nd ACM Int'l. Workshop on Performance Evaluation of Wireless Ad Hoc, Sensor, and Ubiquitous Networks (PE-WASUN) (October 2005)
8. Isaacman, S., et al.: A Tale of Two Cities. In: 11th Workshop on Mobile Computing Systems and Applications (HotMobile), Annapolis, Maryland (February 2010)
9. Isaacman, S., et al.: Ranges of Human Mobility in Los Angeles and New York. In: 8th Int'l Workshop on Managing Ubiquitous Comm. and Services (March 2011)
10. Coluccia, A., et al.: On robust estimation of network-wide packet loss in 3g cellular networks. In: 5th IEEE Broadband Wireless Access Workshop (BWA 2009) (December 2009)
11. Coluccia, A., Ricciato, F., Romirer-Maierhofer, P.: Bayesian Estimation of Network-Wide Mean Failure Probability in 3G Cellular Networks. In: Hummel, K.A., Hlavacs, H., Gansterer, W. (eds.) PERFORM 2010 Workshop. LNCS, vol. 6821, pp. 167–178. Springer, Heidelberg (2011)
12. D'Alconzo, A., et al.: A distribution-based approach to anomaly detection for 3g mobile networks. In: IEEE GLOBECOM (December 2009)
13. Trestian, I., et al.: Measuring serendipity: connecting people, locations and interests in a mobile 3g network. In: Internet Measurement Conference (IMC 2009), Chicago, IL (November 2009)
14. Trestian, I., et al.: Taming user-generated content in mobile networks via drop zones. In: IEEE INFOCOM 2011, Shanghai, China (April 2011)
15. Tanimoto: Internal report. Technical report, IBM (November 1957)
16. TomTom Traffic, http://www.tomtom.com/landing_pages/traffic_solutions/
17. Cellint Traffic Solutions, http://www.cellint.com
18. Decell Technologies Ltd., http://www.decell.com
19. Poolsawat, A., et al.: Acquiring road traffic information through mobile phones. In: 8th Int'l Conference on ITS Telecommunications (October 2008)
20. Minh, Q.T., Kamioka, E.: Granular quantifying traffic states using mobile probes. In: IEEE VTC 2010-Fall (October 2010)
21. Sricharan, M.S., Vaidehi, V.: A Pragmatic Analysis of User Mobility Patterns in Macrocellular Wireless Networks. In: World of Wireless, Mobile and Multimedia Networks (WoWMoM), Espoo, Finland (June 2007)
22. Mobile Millennium, Project, http://traffic.berkeley.edu

23. Hummel, K.A., Hess, A.: Estimating Human Movement Activities for Opportunistic Networking: A Study of Movement Features. In: IEEE International Symposium on a World of Wireless, Mobile and Multimedia Networks (WoWMoM 2011), Lucca, Italy, pp. 1–7 (June 2011)
24. Traffic.online, http://www.its-munich.de/pdf/DynInfoBY/4-Birle.pdf
25. Do-iT T-Mobile D1
26. Gundlegard, D., Karlsson, J.M.: Route Classification in Travel Time Estimation Based on Cellular Network Signaling. In: 12th Int'l IEEE Conference on ITS (2009)
27. Gundlegard, D., Karlsson, J.M.: Handover location accuracy for travel time estimation in GSM and UMTS. In: IET Intelligent Transport System (March 2008)

Identifying Skype Nodes in the Network Exploiting Mutual Contacts[*]

Jan Jusko[1,2] and Martin Rehak[1,2]

[1] Faculty of Electrical Engineering, Czech Technical University in Prague
{jan.jusko,martin.rehak}@fel.cvut.cz
[2] Cognitive-Security s.r.o., Prague, Czech Republic

Abstract. In this paper we present an algorithm that is able to progressively discover nodes of a Skype overlay P2P network. Most notably, super nodes in the network core. Starting from a single, known Skype node, we can easily identify other Skype nodes in the network, through the analysis of widely available and standardized IPFIX (NetFlow) data. Instead of relying on the analysis of content characteristics or packet properties of the flow itself, we monitor connections of known Skype nodes in the network and then progressively discover the other nodes through the analysis of their mutual contacts.

1 Introduction

As recently shown in [7], P2P traffic can degrade the performance of anomaly detection techniques. The detection rate can decrease by up to 30% and false positive rate can increase by up to 45%. In this paper we deal with Skype; a VoIP application based on a P2P protocol. Skype is the perfect example of the VoIP-telephony popularity these days. In March 2011 Skype recorded more than 30 million online users, as reported in [8].

In this paper we focus on detecting Skype nodes in the network using only NetFlow data. Our approach is rather different from those presented so far; Instead of packet inspection and flow analysis [5,6,2] we focus on reconstructing the Skype overlay network from a single Skype node, further called the *seed*.

A description of the Skype architecture and the analysis of its message workflow can be found in [5]. The authors have also created several signatures to detect Skype traffic, based on these findings. Detection using these signatures requires packet payload inspection. In [6] the authors also provide an analysis of user-behavior in the Skype network and estimate the network workload generated by Skype.

One of the first works in the field of Skype detection [3] proposes two detection methods — one based on Pearson's Chi Square test and the second using Naive

---

[*] This material is based upon work supported by the ITC-A of the US Army under Contract W911NF-12-1-0028 and by ONR Global under the Department of the Navy Grant N62909-11-1-7036. Also supported by Czech Ministry of Education grant MSMT ME10051 and MVCR Grant number VG2VS/189.

A. Pescapè, L. Salgarelli, and X. Dimitropoulos (Eds.): TMA 2012, LNCS 7189, pp. 81–84, 2012.

```
function processFlow(flow, Set skypes, Set candidates, Map contacts)
   if (isInteresting(flow))
      local = localEndpoint(flow);
      remote = remoteEndpoint(flow);
      if (skypes.contains(local) || candidates.contains(local))
         contacts.get(local).addOrRenew(remote);
      else if (contacts.anySetContains(remote))
         candidates.add(local);
         contacts.get(local).addOrRenew(remote);

function updateModel(Set skypes, Set candidates, Map contacts)
   contacts.removeOldInactive();
   for (candidate : candidates)
      if (getMaxOverlap(candidate, skypes) > THRESHOLD)
         candidates.remove(candidate);
         skypes.add(candidate);
   skype.removeOldInactive();
```

Fig. 1. Detection algorithm overview. It is split into two routines, the former is used for processing recorded flows, the latter for promoting candidate nodes and cleaning dead connections.

Bayes Classifiers. Since both methods require packet payload inspection their deployment on backbone networks can be problematic. An approach similar to ours was used in [4] to detect members of a P2P botnet.

2   Detection Method

Skype is a P2P application [6] and communicates with other nodes in its overlay network when user signs in (contacting up to 22 distinct nodes [2]) and also when searching for other users. Most of these connections are done over UDP protocol. On average, a Skype node makes 16 connections to other nodes every 5 minutes. The number of distinct contacted nodes rising linearly with time [9]. We use this behavior to reconstruct the Skype overlay network.

The detection algorithm requires at least one known Skype node in order to work. It keeps two data sets in memory - *Skype Nodes Set (SNS)* and *Candidate Nodes Set (CNS)*. A *candidate node* is an endpoint from a local network which has at least one mutual contact with a Skype node. A *contact* (of a node) is an endpoint from outside the protected network that communicates with the given node. If two nodes have the same contact, it is called a *mutual contact*. The algorithm monitors all contacts of each node in SNS and CNS, and stores them in memory for time t. If number of mutual contacts of a candidate node with any Skype node exceeds the threshold k, then the candidate node is labeled as a Skype node and moved to the *SNS*, which contains active Skype nodes. The algorithm also continually monitors traffic of local endpoints which are not contained in either set, whereby if a connection to a contact of any Skype node appears, the given endpoint is added to the *CNS*. All contacts or nodes that have been inactive for at least two hours are removed from memory. The detection algorithm is described in Fig. 1.

Table 1. Percentage of Skype nodes that would be detected if all other Skype nodes were already known for various choices of k and t in *ds1*

k \t	5	15	30	60	90	120	150	180	210	240	∞
1	24.92	54.95	82.28	95.2	96.7	97.6	98.5	98.5	98.5	98.8	99.4
2	7.81	26.13	48.05	81.68	93.39	95.2	95.8	96.1	97	97.3	98.5
3	5.11	13.51	29.43	61.26	79.58	88.59	92.79	95.2	95.5	95.8	98.2
4	2.1	8.41	18.02	43.24	63.96	77.48	85.59	89.49	93.09	94.29	97.9

3   Evaluation

For evaluation of our experiment we used three data sets. The first data set (*ds1*) contains only Skype UDP signaling traffic and is available at [1]. This data set was also used for evaluation in many other works. The second data set (*ds2*) contains 24 hosts managed by ourselves, in the University network, 18 of them running a Skype client. The last data set (*ds3*) was collected on a part of the University network which contains approximately 900 hosts. We do not have access to all computers in this network, therefore we verify detection results manualy or using *nmap*. A detected node is considered to be a true positive if *nmap* recognizes the Skype service or if we manually determine that the endpoint is connecting to one of the bootstrap Skype nodes. We are unable to determine ground truth for all endpoints due to the University's network policies.

All data in *ds1* is generated by Skype, thus in the first experiment we determine what percentage of Skype nodes would be detected by our algorithm if all other Skype nodes are already known. The results presented in Table 1 show that an increase in t improves the detection performance whereby an increase in k worsen it. Furthermore, we can observe that increase of the detection performance with the increase of the stablilizes at some value of t (which happens sooner when choosing lower k values). Moreover, we can get similar detection results for all k values just by increasing t appropriately. Furthermore, we believe that setting k to higher values leads to lower false positive rates, since the probability that two endpoints running different services having x mutual contacts decreases with x.

For *ds2* we estimate both the detection rate and the false positive rate. For each choice of k there is a minimal value of t the algorithm needs to achieve 100% detection rate. Tests using *ds2* managed to achieve 100% detection rate using k as high as 7 and t not greater than 60 minutes (we did not conduct experiments with higher values of k). In this experiment the algorithm managed to keep the false positive rate at 0%.

Regarding the *ds3*, as we can not determine the ground truth for all endpoints, therefore we estimate only false positive rate for this data set (assuming 100% detection rate). We recognize two types of false positive rates — *FPR1* is calculated on the set of endpoints, *FPR2* is determined on the set of flows.

Both rates are very low — with PR1 lower than 0.01% and FPR2 lower than 0.6% for parameters. We were able to divide false positives into three groups:

gateways - endpoints created by a router or host serving as a gateway for a NAT routing Skype traffic, *former Skype nodes* - endpoints that were used by Skype but are no longer active and *Skype's ephemeral ports* - endpoints used by Skype for TCP communication.

Other methods detect packets and flows that represent Skype traffic based on various statistical properties and this detection is instant. We, on the other hand, identify endpoints on which Skype listens for communication. We were therefore interested how long it takes us to detect a Skype node in the network. Our measurements on *ds1* show that out of 18 Skype nodes, 13 were detected within 5 minutes, 4 more within 10 minutes and the final node was detected within one hour. We find this delay acceptable for the purpose of Skype detection.

4   Conclusion

In this paper we present a novel approach to detecting Skype nodes in the network through the reconstruction of a Skype overlay network from a single seed node. This approach exploits the fact that if there are more Skype nodes in the network, then the endpoints they communicate with tend to overlap. Our detection method managed to achieve a high detection rate while keeping the false positive rate to a minimum.

References

1. Skype Traces, `http://tstat.tlc.polito.it/traces-skype.shtml` (acc. May 11, 2011)
2. Baset, S.A., Schulzrinne, H.G.: An Analysis of the Skype Peer-to-Peer Internet Telephony Protocol. In: Proceedings of 25th IEEE International Conference on Computer Communications, IEEE INFOCOM 2006, pp. 1–11. IEEE (2006)
3. Bonfiglio, D., Mellia, M., Meo, M., Rossi, D., Tofanelli, P.: Revealing skype traffic: when randomness plays with you. ACM SIGCOMM Computer Communication Review 37(4), 37–48 (2007)
4. Coskun, B., Dietrich, S., Memon, N.: Friends of an enemy: identifying local members of peer-to-peer botnets using mutual contacts. In: Proceedings of the 26th Annual Computer Security Applications Conference, ACSAC 2010, pp. 131–140. ACM, New York (2010)
5. Ehlert, S., Petgang, S., Magedanz, T.: Analysis and signature of Skype VoIP session traffic. In: 4th IASTED International (2006)
6. Guha, S., Daswani, N., Jain, R.: An experimental study of the skype peer-to-peer voip system. In: Proceedings of IPTPS, vol. 6, pp. 5–10. Citeseer (2006)
7. Haq, I.U., Ali, S., Khan, H., Khayam, S.A.: What Is the Impact of P2P Traffic on Anomaly Detection? In: Jha, S., Sommer, R., Kreibich, C. (eds.) RAID 2010. LNCS, vol. 6307, pp. 1–17. Springer, Heidelberg (2010)
8. Parkes, P.: 30 million people online on Skype (2011), `http://blogs.skype.com/en/2011/03/30_million_people_online.html` (acc. August 24, 2011)
9. Rossi, D., Mellia, M., Meo, M.: Following skype signaling footsteps. In: 2008 4th International Telecommunication Networking Workshop on QoS in Multiservice IP Networks, pp. 248–253. IEEE (February 2008)

Padding and Fragmentation for Masking Packet Length Statistics

Alfonso Iacovazzi and Andrea Baiocchi

Dept. of Information Engineering, Electronics and Telecommunications (DIET),
University of Roma Sapienza, Via Eudossiana 18, 00184 Roma, Italy
iacovazzi@infocom.uniroma1.it, andrea.baiocchi@uniroma1.it

Abstract. We aim at understanding if and how complex it is to ob-
fuscate traffic features exploited by statistical traffic flow classification
tools. We address packet length masking and define perfect masking as
an optimization problem, aiming at minimizing overhead. An explicit
efficient algorithm is given to compute the optimum masking sequence.
Numerical results are provided, based on measured traffic traces. We find
that fragmenting requires about the same overhead as padding does.

1 Introduction

In this work we investigate *protection* of privacy against traffic analysis (see
[1][2]) and, at the same time, how much effort is to be devoted to fool traffic
analysis tools. Some recent works [4] [5] have made a careful analysis of the information conveyed by the various features used to classify different application
classes: thier conclusion is that packet lengths are most valuable to traffic classifiers at application level. Moreover, in [3] Authors prove experimentally they
are capable of getting partial transcript of encrypted VoIP transactions, by exploiting also statistical information leaked by message lengths. This is the reason
why we investigate how packet length information can be concealed to statistical
traffic analysis, besides ciphering. We term this *packet length masking*.

In [6] the Authors propose a technique for changing the packet lengths by
optimally morphing one class of traffic to look like another class. They make
use of convex optimization techniques to modify the packet lengths in order to
get the minimum introduced overhead; however they do not account for packet
fragmentation overhead, for residual correlations among masked packet lengths.
Shui Yu et al. [7] implement a strategy to introduce dummy packet padding into
a flow in order to guarantee perfect anonymity on web browsing. They replace
dummy packets with prefetched data in order to solve problems of extra delay
and additional cost of bandwidth due to the packet padding.

We define formally the traffic classification problem assess the achievable performance bounds for a masking algorithm, by defining the optimal solution to
the stated masking problem. We show with many numerical examples based on
real traffic traces that fragmenting entails a marginal benefit as to overhead with
respect to much simpler approaches where only padding is used.

A. Pescapè, L. Salgarelli, and X. Dimitropoulos (Eds.): TMA 2012, LNCS 7189, pp. 85–88, 2012.
© Springer-Verlag Berlin Heidelberg 2012

2   Mixing Two Applications: Optimal Solution

We consider a packet network where we can identify two hosts running a given application protocol within a secure (ciphered) communication channel. Let us consider an application flow and let $\mathbf{X} = [X_1, \ldots, X_m]$ be a sequence of positive, discrete random variables modeling the lengths of the first m packets of application $\mathcal{A}_j$ flows, with $X_k^{(j)} \in [1, \ell]$, where ℓ is the maximum packet length. We consider two application, $\mathcal{A}_0$ and $\mathcal{A}_1$. Assume the sample space of application $\mathcal{A}_i$ be $\Omega_i \subset [1, \ell]^m$, with $\omega_i = |\Omega_i|$, for $i = 0, 1$. Application $\mathcal{A}_j$ is characterized by a specific probability distribution function of the random vector $\mathbf{X}^{(j)}$, denoted as $p_j(\mathbf{x}) = \mathcal{P}(\mathbf{X}^{(j)} = \mathbf{x})$, where $\mathbf{x}$ is a vector of positive integers not greater than ℓ.

Padding and fragmenting functions substitute the i-th packet of a flow with n_i packets carrying the original information bytes, plus overhead. In terms of length values, the single value $X_i = H + L_i$ is turned into the length values $Y_{i,r} = H + L_{i,r} + U_{i,r}$ for $r = 1, \ldots, n_i$, with $L_{i,1} + \cdots + L_{i,n_i} = L_i$ and $U_{i,r} \geq 0$. We denote this transformation briefly with $\mathbf{Y} = \phi(\mathbf{X})$. The adversary observes the sequence $\{Y_k\}_{1 \leq k \leq n}$, with $n = n_1 + \cdots + n_m$, and knows any algorithm the sender may have used to pad, fragment and encipher the packets of the original flow with lengths $\mathbf{X}$. The aim of the adversary is to guess the application the original flow belongs to. This is summarized by an algorithm named $TA(\mathbf{Y})$. The best that the masking algorithm can do is to devise the transformation $\mathbf{Y} = \phi(\mathbf{X})$ so that, whatever $p_0(\mathbf{x})$ and $p_1(\mathbf{x})$ be, it is $q_0(\mathbf{y}) = q_1(\mathbf{y})$, where $q_i(\cdot)$ is the probability distribution function of $\mathbf{Y}^{(i)} = \phi(\mathbf{X}^{(i)})$, $i = 0, 1$.

To construct the transformation $\phi(\cdot)$ we define ordered couples $(\mathbf{x}_h^{(0)}, \mathbf{x}_k^{(1)})$, with $\mathbf{x}_h^{(0)} \in \Omega_0$ and $\mathbf{x}_k^{(1)} \in \Omega_1$, for $h = 1, \ldots, \omega_0$ and $k = 1, \ldots, \omega_1$. For each couple $(\mathbf{x}_h^{(0)}, \mathbf{x}_k^{(1)})$ we find the optimum flow pattern $\mathbf{y}_{h,k}$ made up of $n_{h,k}$ packet lengths that the flows in the couple can be mapped to by means of padding and fragmentation. Optimum here refers to minimization of the overhead required to convert each of the two flows of the couple into the masked flow $\mathbf{y}_{h,k}$. Since the packet lengths of the input flows are given, we can take as optimization target the function $E_{h,k} = |\mathbf{y}_{h,k}| + n_{h,k}H$, that is simply the overall length of the output masked flow $\mathbf{y}_{h,k}$. Perfect masking is obtained by requiring that $\mathcal{P}(\mathbf{Y} = \mathbf{y}_{h,k} | \mathbf{X} = \mathbf{x}_h^{(0)}) \cdot p_0(\mathbf{x}_h^{(0)}) = \mathcal{P}(\mathbf{Y} = \mathbf{y}_{h,k} | \mathbf{X} = \mathbf{x}_k^{(1)}) \cdot p_1(\mathbf{x}_k^{(1)}) \equiv c_{h,k}$. This requirement guarantees that the adversary has no clue to what application has emitted the original flow from the observation of the masked flow $\mathbf{y}_{h,k}$. Then the problem is to determine $c_{h,k}$ so as to minimize $z = \sum_{h=1}^{\omega_0} \sum_{k=1}^{\omega_1} c_{h,k} \cdot E_{h,k}$ subject to constraints $0 \leq c_{h,k} \leq \min\left\{p_0(\mathbf{x}_h^{(0)}), p_1(\mathbf{x}_k^{(1)})\right\}$, $\sum_{k=1}^{\omega_1} c_{h,k} = p_0(\mathbf{x}_h^{(0)})$, $\sum_{h=1}^{\omega_0} c_{h,k} = p_1(\mathbf{x}_k^{(1)})$. This optimization problem is related to the well-known Transportation Problem [8], with the only difference that in our case we have the quantities to "transport" expressed as fractions of the total amount. The problem at hand is solved by Hungarian algorithm [9].

The quantities $E_{h,k}$ represent the minimum amount of overhead required to mix the h-th flow of application $\mathcal{A}_0$ with the k-th one of application $\mathcal{A}_1$. If we consider just padding and then no fragmentation, $E_{h,k}$ can be easily computed through the relation $E_{h,k} = \sum_{r=1}^{m} |x_h^{(0)}(r) - x_k^{(1)}(r)|$. If fragmentation is used as well, computing the weights $E_{h,k}$ is more complex, but it can be done simply by exhaustive search among all flow patterns $\mathbf{y}_{h,k}$ derived via padding and fragmentation from $\mathbf{x}_h^{(0)}$ and $\mathbf{x}_k^{(1)}$, provided m is small, e.g. less than 10.

The ideal masking algorithm for two applications is:

1. take as input a flow $\varphi_{h^*} \in \mathcal{A}_0$ (or: $\psi_{k^*} \in \mathcal{A}_1$);
2. draw a random index in the set $[1, \omega_1]$ of value k^* with probability $\frac{c_{h^*,k^*}}{p_0(\varphi_{h^*})}$
   (or: in the set $[1, \omega_0]$ of value h^* with probability $\frac{c_{h^*,k^*}}{p_1(\psi_{k^*})}$);
3. transform the input flow φ_{h^*} (or: ψ_{k^*}) into the output masked flow $\mathbf{y}_{h^*,k^*}$.

We have implemented this algorithm by filling the table of output flows $\mathbf{y}_{h,k}$ and solving the above stated optimization problem to find the values of $c_{h,k}$.

3   Numerical Results and Discussion

Performance of the algorithms developed in this work is analyzed by using the same datasets collected and commented in [10]. The traffic is generated by considering clients and servers placed in University lab and industrial lab LANs, as well as private domestic locations. All connections are through public Internet. Servers and clients use different operating systems, so as to collect a representative traffic sample. Collected flows belong to different application protocols: HTTP, SSH, POP3, FTP (control session, FTP-c), VoIP. For each of the considered application protocols 2000 flows are collected.

We have compared the optimum masking algorithm for two application applied in two different ways. In the first case, we apply both fragmentation and padding when constructing optimum output flow $\mathbf{y}_{h,k}$ paired with input flows $\mathbf{x}_h^{(0)}$ and $\mathbf{x}_k^{(1)}$. In the second case, only padding is allowed, so that $\mathbf{y}_{h,k}$ is made up m packets, whose lengths are just $y_{h,k}(r) = \max\{x_h^{(0)}(r), x_k^{(1)}(r)\}$, $r = 1, \ldots, m$. All considered approaches lead to full masking of traffic flows as regards the packet length information.

We have considered various couples of applications drawn from the HTTP, SSH, POP3, FTP-c and VoIP. In the table 1 we compare the average overhead required by the different masking algorithms. Fixed-size packets is reported in the third column; padding only optimum masking is given in the second column; joint padding and fragmentation optimum masking is given in the first column.

Table 1. Average amount of overhead introduced by different packet length masking algorithms for various couple of application flows

Applications Pair	Fragmentation AND Padding	Padding Only	Applications Pair	Fragmentation AND Padding	Padding Only
HTTP - SSH	0.3191	0.3605	SSH - VoIP	0.2353	0.3546
HTTP - FTP-c	0.4088	0.4126	FTP-c - POP3	0.2094	0.2336
POP3 - HTTP	0.4351	0.4392	FTP-c - VoIP	0.2303	0.2752
HTTP - VoIP	0.3864	0.4126	POP3 - VoIP	0.2477	0.2477
SSH - FTP-c	0.3008	0.3162	HTTP over SSH - SFTP	0.2090	0.2248
SSH - POP3	0.3495	0.3715			

We notice that the amount of overhead can vary significantly depending on the applications we mix. In addition, we can see that masking with fixed-size packets leads to introduction of high amount of overhead, while algorithm with only padding does not cause a significant increase in overhead compared to the generalized masking in which we can fragment. In a lot of cases, adding fragmentation improve marginally the achieved fraction of overhead. This is a strong argument advocating the use of padding only, although this is not intuitive at first. Main reason is that full masking requires not only masking the length of each packets of the flow but also the amount of bytes of the entire flow.

The aim of the study is to assess what the optimum performance in terms of overhead could be under the constraint of perfect masking. The approach taken assumes the entire sequence of packet lengths of a flow used in the masking process can be observed to take the masking decision and joint statistics of packet lengths are known. Actual application usually develop through a bidirectional message exchange, made of requests and replies, which makes our approach non practical. It servers the purpose to establish a theoretical bound. We are working towards practical solutions that mask flow features (lengths, inter-packet times, directions) and operate packet by packet, require only marginal packet length statics and allow relaxation of the full masking constraint.

References

1. Callado, A., Kamienski, C., Szabo, G., Gero, B., Kelner, J., Fernandes, S., Sadok, D.: A Survey on Internet Traffic Identification. IEEE Communications Surveys & Tutorials 11(3), 37–52 (2009)
2. Kim, H., Claffy, K., Fomenkov, M., Barman, D., Faloutsos, M., Lee, K.: Internet traffic classification demystified: myths, caveats, and the best practices. In: Proc. of ACM CoNEXT, Madrid, Spain, December 9-12 (2008)
3. White, A.M., Matthews, A.R., Snow, K.Z., Monrose, F.: Phonotactic Reconstruction of Encrypted VoIP Conversations: Hookt on fon-iks. In: Proc. of the 32nd IEEE Symposium on Security and Privacy, Berkeley, CA, USA, May 22-25 (2011)
4. Este, A., Gringoli, F., Salgarelli, L.: On the stability of the information carried by traffic flow features at the packet level. ACM SIGCOMM Computer Communication Review 39(3) (2009)
5. Lim, Y., Kim, H., Jeong, J.: Internet Traffic Classification Demystified: On the Sources of the Discriminative Power. In: Proc. of ACM CoNEXT, Philadelphia, USA (2010)
6. Wright, C.V., Coull, S.E., Monrose, F.: Traffic Morphing: An Efficient Defense Against Statistical Traffic Analysis. In: Proc. of the 16th Network and Distributed System Security Symposium (NDSS), San Diego, CA, USA, February 8-11 (2009)
7. Yu, S., Thapngam, T., Wei, S., Zhou, W.: Efficient Web Browsing with Perfect Anonymity Using Page Prefetching. In: Hsu, C.-H., Yang, L.T., Park, J.H., Yeo, S.-S. (eds.) ICA3PP 2010, Part I. LNCS, vol. 6081, pp. 1–12. Springer, Heidelberg (2010)
8. Hitchcock, F.L.: The distribution of a product from several sources to numerous localities. J. Math. Phys. 20, 224–230 (1941)
9. Kuhn, H.W.: The Hungarian method for the assignment problem. Naval Research Logistics Quarterly 2, 83–97 (1955)
10. Maiolini, G., Molina, G., Baiocchi, A., Rizzi, A.: On the fly Application Flows Identification by exploiting K-Means based classifiers. Journal of Information Assurance and Security (2), 142–150 (2009)

Real-Time Traffic Classification Based on Cosine Similarity Using Sub-application Vectors*

Cihangir Beşiktaş[1,2] and Hacı Ali Mantar[1]

[1] Department of Computer Engineering, Gebze Institute of Technology
cbesiktas@gyte.edu.tr,hamantar@bilmuh.gyte.edu.tr
[2] TUBITAK BILGEM, Center of Research for Advanced Technologies
of Informatics and Information Security
cihangirbesiktas@uekae.tubitak.gov.tr

Abstract. Internet traffic classification has a critical role on network monitoring, quality of service, intrusion detection, network security and trend analysis. The conventional port-based method is ineffective due to dynamic port usage and masquerading techniques. Besides, payload-based method suffers from heavy load and encryption. Due to these facts, machine learning based statistical approaches have become the new trend for the network measurement community. In this short paper, we propose a new statistical approach based on DBSCAN clustering and weighted cosine similarity. Our experimental test results show that the proposed approach achieves very high accuracy.

Keywords: Traffic classification, cosine similarity, DBSCAN, packet inspection.

1   Introduction

Real-time Internet traffic classification provides intrusion detection, firewalling, quality of service, trend analysis, internet monitoring and lawful interception functionalities. These essential services of traffic classification have increased the importance of research in this area.

The current Internet traffic classification proposals find out the applications of network flows based on five tuples IP header(source and destination IPs, source and destination ports and transport layer protocol). Some of them are based on port numbers and L7 payloads. However, dynamic or registered port usage, encryption and heavy load cause these approaches being ineffective. Due to these facts, machine learning based statistical approaches have become the new trend in this research field.

A survey of ML techniques can be viewed in [1]. Besides this paper drew inspiration from the study of Chung et al.[2] which proposes a cosine similarity based approach using payload vectors. But since they focus on the content of packet payloads, their work is supposed to be ineffective for encrypted traffic.

---
* This work has been supported by Inforcept Networks corporation.

A. Pescapè, L. Salgarelli, and X. Dimitropoulos (Eds.): TMA 2012, LNCS 7189, pp. 89–92, 2012.
© Springer-Verlag Berlin Heidelberg 2012

This paper presents a new ML based statistical methodology that extracts sub-application vectors of each application using DBSCAN clustering algorithm in the learning phase. The methodology uses weighted cosine similarity in the classification phase.

2 Methodology

Our classifier has two main components: off-line training and on-line classification. These components are presented in the following sections.

2.1 Off-line Training

The evaluation of each application's characteristic features constitutes our ML method's training phase. This phase is done with previously collected training dataset in offline mode. The following lists the steps of the training phase.

1. Our port and payload based pre-classifier detects the application of each flow in the training dataset, and records each flow's signed packet sizes of first few packets to the detected application's dataset. Here upstream packet which flows from client to server has minus sign.
2. Each application's dataset is divided into clusters using DBSCAN algorithm to calculate sub-application vectors of the application. Each cluster has minpts number of flows which are connected together in an epsilon distance which is measured using cosine similarity.
3. For each cluster, sub-application vectors which consist of signed packet sizes vector and standard deviation vector are computed.

We have selected DBSCAN clustering due to the observations that it produces variable length clusters which have tightly connected nodes in [3]. By this way, the characteristics of applications are computed more precise. DBSCAN clustering can be viewed in [4].

After DBSCAN clustering, sub-application vectors of a given cluster can be mathematically defined as follows.

Assume we have a cluster C which has N number of flows, and each flow in the cluster has signed packet sizes vector. If the jth signed packet size of the ith flow of the cluster is shown as Sz_{ji}, then the ith element of the signed packet size vector of the cluster can be calculated using Equation 1.

$$SPSz(i) = \left(\frac{\sum_{j=1}^{N} Sz_{ji}}{N} \right) \tag{1}$$

After that, the ith element of the normalized standard deviation vector can be observed using Equation 2 and Equation 3.

$$Stdev(i) = \left(\sqrt{\frac{\sum_{j=1}^{N} (Sz_{ji} - SPSz(i))^2}{N}} \right) \tag{2}$$

$$NStdev(i) = \frac{1}{1 + Stdev(i)} \qquad (3)$$

The idea behind the normalization of standard deviation is to bounce it between 1 and 0. Hence, standard deviations which are close to zero are limited to 1 to have higher weights in weighted cosine similarity measurement.

2.2   On-line Classification

After the evaluation of each application's characteristic features, detecting the application type of an unknown flow forms the classification phase of our ML based approach. In order to classify an unknown flow, cosine similarities between the unknown flow's packet size vector and all the packet size vectors of applications are calculated, and the application with the highest value is assigned as the application type of the unknown flow. Cosine similarity can be viewed in [5].

3   Experimental Results

In this section, we present the experimental results of our approach. We have tested our approach using four different methods to show the effects of sub-signature vectors and standard deviation. These four methods are created based on their usage of sub-application vectors and standard deviation.

The first method which uses just packet sizes vectors is our base point. The second method which uses packet sizes vectors and standard deviation vectors is to obtain the impact of standard deviation vectors. The third method which uses packet sizes vectors with sub-application vectors shows the effect of sub-application vectors. The last method which uses both standard deviation vectors and sub-application vectors is to see the influence of both.

3.1   Implementation and Datasets

We have implemented our classifier in FreeBSD 8.2 operating system using C language and libpcap packet capture library.

We have collected two full-payload traces to evaluate the performance of our methodology. The first trace (Trace 1) was collected at an edge network which serves 10 hosts. The second trace (Trace 2) was collected at a Turkish telecom operator's 3G backbone which has 70 MByte/s throughput.

3.2   Test Results and Discussion

This section reviews and discusses the results of the four methods mentioned above for Trace 1 and Trace 2.

The accuracy results which are the percentage of true positives are given in Table 1. For Trace 1, It is obvious that method 1 which uses just packet sizes, and method 2 which uses packet sizes and standard deviations of flows perform very low accuracy. However, it can be inferred from the results that standard deviation increases the accuracy. Moreover, method 3 which uses sub-application vectors and method 4 which uses sub-application vectors with standard deviation achieve spectacular increase of the accuracy, and their performances are close to each others with method 4 leading a few steps ahead. For Trace 2, similar observations can be made. This shows that our methodology performs well both in edge networks and backbones.

Table 1. Results for Trace 1 and Trace 2

Trace:	Trace 1				Trace 2			
Method:	Method 1	Method 2	Method 3	Method 4	Method 1	Method 2	Method 3	Method 4
HTTP	0.13	0.19	0.95	0.94	0.47	0.54	0.89	0.92
HTTPS	0.23	0.14	0.90	0.70	0.28	0.57	0.97	0.96
FTP	0.22	0.41	0.997	0.996	0.0	0.0	0.84	0.84
DNS	0.42	0.92	1.0	1.0	0.80	0.82	0.98	0.94
Bittorrent	-	-	-	-	0.35	0.40	0.74	0.63
Edonkey	-	-	-	-	0.82	0.83	0.75	0.82
MSN	-	-	-	-	0.20	0.68	0.93	0.96
XMPP	-	-	-	-	0.68	0.68	0.94	0.91
ESMTP	-	-	-	-	0.81	0.83	0.93	0.93
POP3	-	-	-	-	0.28	0.30	0.87	0.91

4   Conclusion

In this paper, we proposed an ML based statistical classification technique based on DBSCAN clustering and cosine similarity. Our feature set contains only signed packet sizes of first few packets and their standard deviations which make our approach have the ability to classify encrypted applications.

Our experimental analysis shows that our contributions sub-application vectors and standard deviation achieve high accuracy.

References

1. Nguyen, T.T.T., Armitage, G.: A Survey of Techniques for Internet Traffic Classification using Machine Learning. IEEE Communications Surveys & Tutorials 10(4), 56–76 (2008)
2. Chung, J.Y., Park, B., Won, Y.J., Strassner, J., Hong, J.W.: Traffic Classification Based on Flow Similarity. In: Nunzi, G., Scoglio, C., Li, X. (eds.) IPOM 2009. LNCS, vol. 5843, pp. 65–77. Springer, Heidelberg (2009)
3. Erman, J., Arlitt, M., Mahanti, A.: Traffic Classification Using Clustering Algorithms. In: SIGCOMM 2006 Workshops, Pisa, Italy (2006)
4. DBSCAN Wikipedia, http://en.wikipedia.org/wiki/DBSCAN
5. Cosine Similarity Wikipedia, http://en.wikipedia.org/wiki/Cosine_similarity

Towards Efficient Flow Sampling Technique for Anomaly Detection

Karel Bartos and Martin Rehak

Faculty of Electrical Engineering
Czech Technical University in Prague
Prague, Czech Republic
{karel.bartos,martin.rehak}@agents.fel.cvut.cz

Abstract. With increasing amount of network traffic, sampling techniques have become widely employed allowing monitoring and analysis of high-speed network links. Despite of all benefits, sampling methods negatively influence the accuracy of anomaly detection techniques and other subsequent processing. In this paper, we present an adaptive, feature-aware sampling technique that reduces the loss of information bounded with the sampling process, thus minimizing the decrease of anomaly detection efficiency.

To verify the optimality of our proposed technique, we build a model of the ideal sampling algorithm and define general metrics allowing us to compute the distortion of traffic feature distribution for various types of sampling algorithms. We compare our technique with random flow sampling and reveal their impact on several anomaly detection methods by using real network traffic data. The presented ideas can be applied on high-speed network links to refine the input data by suppressing highly-redundant information.

Keywords: sampling, anomaly detection, NetFlow, intrusion detection.

1 Introduction

The increasing amount of the network traffic related to the growing speeds of the modern high-speed networks implicates the need of more efficient and powerful monitoring devices. One possibility, how to deal with these requirements, is to use sampling methods. This field was studied in detail during last years. Various approaches and methods were published presenting how to sample the traffic, but the main interest was focused on the preservation of the volume characteristics of the monitored traffic such as the number of transferred bytes, packets and flows. The impact of the sampling methods on the anomaly detection systems was out of the interest of the network researchers till the last several years.

The use of sampled data for more advanced analysis, such as Network Behavior Analysis, is problematic [14], as the sampling severely harms the effectiveness of the anomaly detection and data analysis algorithms. These algorithms are based on pattern recognition and statistical traffic analysis, and the distortion of traffic features can significantly increase the error rate of these underlying methods by breaking their assumptions about traffic characteristics.

A. Pescapè, L. Salgarelli, and X. Dimitropoulos (Eds.): TMA 2012, LNCS 7189, pp. 93–106, 2012.
© Springer-Verlag Berlin Heidelberg 2012

In this paper, we build a formal model allowing us to analyze the impact of sampling on traffic feature distributions (Section 3). In Section 4, we introduce the combination of adaptive sampling algorithm and late sampling technique, which optimizes the sampling algorithm behavior w.r.t. the consequent anomaly detection. The improvements are clearly outlined in Section 5, where we compared four types of sampling techniques with the proposed sampling approach.

2 Related Work

There are two basic classes of sampling techniques: packet-based and flow-based methods. *Packet-based sampling* methods work on the level of network packets. Each packet is selected for monitoring with a predefined probability depending on the sampling method used. The main advantage of sampling deployment was the decreased requirements for memory consumption and CPU power on routers as well as the possibility to monitor higher network speeds.

Although packet sampling is easy to implement, it introduces a serious bias in flow statistics [6,11]. An application of packet sampling for traffic analysis, planning and management purposes has been studied in [7,9]. Further research in packet sampling introduced adaptive packet sampling techniques [9], [4]. These techniques adjust the sampling rate depending on the current traffic load in order to acquire more accurate traffic statistics. In the work by [1] is proposed a new type of packet sampling, where malicious packets are sampled with higher probability, which improves the quality of anomaly detection.

In case of *flow sampling*, the monitored traffic is aggregated into network flows and the sampling itself is applied not to the particular packets, but to the whole flows. The main benefit is better accuracy when compared to packet sampling [11], but they require more memory and CPU power.

Smart sampling [6] and sample-and-hold [10] techniques were introduced in order to reduce the memory requirements. Both of these techniques are focused on accurate traffic estimation for larger flows, so called *heavy-hitters*. A combination of packet sampling and flow sampling is presented in [18].

A comparison of packet and flow sampling can be found in [11]. Flow sampling is superior in flow distribution preservation, while the smart sampling prefers large flows over small ones. Comprehensive literature review can be found in [5].

Some recent papers do not focus only on the accuracy of sampling methods, but also on their effect on anomaly detection. The authors of [14] evaluate how the performance of anomaly detection algorithms is affected by selected sampling methods. Their results demonstrate that random packet sampling introduces a measurable bias and decreases the effectiveness of the detection algorithms. Overall, random flow sampling proved to be a good choice. The work by [3] proposes selective flow sampling suitable for specific anomaly detectors. The presented results demonstrate that selective sampling introduces serious bias in traffic feature distribution, which limits its effectiveness and wider usability.

The effect of opportunistic sampling methods (selective flow sampling, smart sampling) on the anomaly detection is studied in [2]. The authors present a

Table 1. The overview of the sampling methods, levels (P - packet, F - flow) and their suitability for anomaly detection and for preserving traffic volumes and distributions. Legend: × - not suitable, ○ - partially suitable, • - suitable.

Sampling Method	Sampling Level	Volume Preserv.	Distrib. Preserv.	Anomaly Detection
Random Packet Sampling	P	×	×	×
Adaptive Packet Sampling	P	○	×	×
Adaptive Non-Linear Sampling	P	•	×	×
Random Flow Sampling	F	•	○	○
Smart Sampling	F	•	×	×
Sample and Hold Sampling	F	•	×	×
Selective Sampling	F	×	×	•

"magnification" of entropy change during particular attacks introduced by opportunistic sampling methods.

The overview of described sampling methods is presented in the Table 1. We can see that the majority of sampling methods were designed for traffic monitoring, while preservation of traffic features crucial for anomaly detection is suboptimal. We shall mention that *random flow sampling* method provides relatively good results in all three areas. *Selective sampling* method is specifically designed for one type of anomaly detection methods.

3   Ideal Flow Sampling

In this Section, we present model of the ideal sampling designed for anomaly detection algorithms. Intuitively, the ideal sampling should be a process in which number of samples and their distribution are selected in such a way that the loss of information is minimal. In other words, by using the ideal sampling it is possible to reconstruct as much of the original information as possible. From this point of view, random sampling satisfies well such requirement.

However, when dealing with the ideal sampling for anomaly detection, different types of information may have different impact on detection performance, simply because of the fact that anomaly detection methods use only some of the information received from the network traffic. This implies that some information is more important (from the network security perspective) than other.

Each flow from network traffic (denoted as x) can be identified by five features (source and destination IP address, source and destination port, and protocol). Besides these basic features, each flow contains additional information like number of bytes or packets transferred, timestamp etc. We will denote k-th feature as X_k. Anomaly detection methods compute statistics of these features serving as an input for their detection procedures. That is why it is reasonable to preserve as much of these statistics as possible.

The statistics can be described by *feature moments* which are computed from feature values. We distinguish between two types of feature moments: *feature counts* $\mathsf{c}(x \mid X_k)$ and *feature entropies* $\mathsf{e}_{X_k}(x \mid X_l)$. Feature counts indicate summations and numbers of flows related to x through the feature X_k (in flows, packets or bytes). We will use feature counts in number of flows unless told otherwise. On the other side, feature entropies describe entropy of feature X_k of flows related to x through the feature X_l. These two types of moments are used in most of existing anomaly detection methods and we consider them as relevant information which the ideal sampling should be able to preserve.

Furthermore we will denote the original finite unsampled set as U and the finite set of samples as S. Thus $c^{\mathsf{U}}(x|sIP)$ denotes the number of flows from the original set with exactly the same source IP address as has this flow x. While $e^{\mathsf{S}}_{sP}(x|dIP)$ is the entropy of source ports from the sampled set, whose flows target exactly the same destination IP address as flow x. When we will consider feature counts across more (q) features, we will denote them as $c(x|X_1, \ldots, X_q)$.

Definition 1. Let S be set of flows selected from the original set U by using probability $p(x)$. We distinguish between two distance functions:

$$d_e(c^{\mathsf{U}}(x|X_k), c^{\mathsf{S}}(x|X_k)) = |c^{\mathsf{U}}(x|X_k) \cdot p(x) - c^{\mathsf{S}}(x|X_k)|,$$
$$d_l(c^{\mathsf{U}}(x|X_k), c^{\mathsf{S}}(x|X_k)) = |c^{\mathsf{U}}(x|X_k) - c^{\mathsf{S}}(x|X_k)|.$$

Both distance functions describe the loss of information for concrete moment values. When the value is equal to zero (and no loss of information is present), it is possible to reconstruct the original moment values - we will call this property as reversibility. Note that the subscript is related to early (e) or late (l) sampling described in Section 4.

Definition 2. Let $\mathsf{S}_1, \ldots, \mathsf{S}_m$ be various sets of flows selected from U. Feature moment $c(x|X_k)$ is reversible in U if and only if:

$$\forall x \in \mathsf{U} : \lim_{m \to \infty} \sum_{i=1}^{m} d(c^{\mathsf{U}}(x|X_k), c^{\mathsf{S}_i}(x|X_k)) = 0,$$

where d is one of the distance functions from Def. 1.

Reversibility ensures complete reconstruction of the original data by using only sampled set, which is key sampling property from the anomaly detection perspective.

To define the reversibility of entropy feature moments, it is reasonable to use relative uncertainty instead of entropy values, because relative uncertainty is normalized and specifies well feature distributions. In the following, we will denote relative uncertainty that describes entropy feature moment $e_{X_k}(x|X_l)$ as:

$$RU(e_{X_k}(x|X_l)) = \frac{e_{X_k}(x|X_l)}{\log c(x|X_l)} \in [0, 1].$$

Definition 3. Let $\mathsf{S}_1, \ldots, \mathsf{S}_m$ be various sets of flows selected from U. Feature moment $e_{X_k}(x|X_l)$ is reversible in U if and only if:

$$\forall x \in \mathsf{U} : \lim_{m \to \infty} \sum_{i=1}^{m} \left(RU(e^{\mathsf{U}}_{X_k}(x|X_l)) - RU(e^{\mathsf{S}_i}_{X_k}(x|X_l)) \right) = 0.$$

Definition 4. Ideal sampling with sampling probability $p(x)$ is defined as a sampling where all feature moments (counts and entropies) are reversible.

Having the ideal sampling defined, we can evaluate quality of different sampling techniques by using the following quality metrics.

Definition 5. We define **feature reversibility** of count moment $c(x|X_k)$, which describes the similarity (interval [0,1]) of a probability distribution with the ideal distribution, and thus measures the reconstruction error, as:

$$f_c^r(X_k) = 1 - \frac{1}{|U|} \cdot \sum_{\forall x \in U} | \ d(c^U(x|X_k), c^S(x|X_k)) \ |$$

and feature reversibility of entropy moment $e_{X_k}(x|X_l)$:

$$f_e^r(X_k, X_l) = 1 - \frac{1}{|U|} \cdot \sum_{\forall x \in U} | \ RU(e_{X_k}^U(x|X_l)) - RU(e_{X_k}^S(x|X_l)) \ |$$

However, high feature moment reversibility does not guarantee the best possible distribution. The case of many missing entities with small moment values may still produce high overall reversibility, yet it may be far from the ideal distribution. From this reason, we use second quality metrics that takes into consideration coverage/variability of values across the feature.

Definition 6. Feature coverage of feature X_i describing the relative variability is defined as:

$$f^c(X_i) = \frac{v(X_i^S)}{v(X_i^U)} \in [0, 1],$$

where $v(X_i^S), v(X_i^U)$ stands for number of distinct values of X_i in S and U.

These two quality metrics caters to conflicting demands in anomaly detection domain – to retain as much of the moment values as possible with minimal loss in precision. High feature moment reversibility is desired in detection techniques based on statistical methods, while maximizing feature variability is essential for knowledge-based approaches that depend on specific values of the individuals.

4   Adaptive Sampling Technique

In this Section, we present an adaptive, feature-aware sampling technique specifically designed for anomaly detection methods. The main ideas behind the proposed method are the following: (1) the incremental value of flows in a single set (defined by one or more common feature values) decreases with the growing number of similar flows already in the set, and (2) the system computes the moments *before* the sampling procedure, so the moment values are computed from the original, full set of data. Before we present the algorithm itself, we first discuss these two assumptions in more detail.

The first idea is based on the assumption that in case of high occurrences of the same feature value in the original set, the benefit of adding this value to the same sampled set decreases in time. In other words, instead of selecting an immense number of the same values n_1, it may be reasonably better to select only satisfactory large subset of these values n_2 ($n_2 < n_1$) without compromising its great magnitude. This way the algorithm is able to save considerable amount of space which can be used for other portions of information.

The second idea of computing feature moments before sampling (we call this procedure as late sampling) advantages of this approach against the traditional technique (called early sampling) in terms of detection quality. Late sampling benefits from the fact that it retains exact statistical information about the original set, both for sampling technique, and for subsequent anomaly detection. To make this method applicable, the computational cost of feature extraction phase should be lower when compared to the cost of the rest of the processing.

4.1   The Algorithm

A majority of anomaly detection methods use values of features in their detection process. The correctness of these key features is vitally important for the overall detection quality – hence we will call them as *primary features*, while the rest of the features are denoted as *secondary features*. Note that a priori knowledge of used anomaly detectors is required for selecting appropriate primary features.

Definition 7. Let $X_1, \ldots, X_k$ be primary features. We define primary probability as the probability that a flow related to x through features $X_1, \ldots, X_k$ is selected to the sampled set:

$$p_p(x|X_1, \ldots, X_k) = \begin{cases} s & \mathsf{c}(x|X_1, \ldots, X_k) \leq t \\ s \cdot \frac{\log t}{\log \mathsf{c}(x|X_1, \ldots, X_k)} & \mathsf{c}(x|X_1, \ldots, X_k) > t \end{cases}$$

where $s \in [0, 1]$ is sampling rate and threshold t defines a point in the distribution, where our sampling technique starts to set the probability proportionally to the size of the moment. The higher the moment value, the lower sampling rate is assigned.

This modification from the random sampling slightly shifts the original probability distribution for flows with moment values above the threshold. We argue that reducing the size of attacks with higher values of moments (above the threshold) that are mostly easily detectable does not harm the anomaly detection effectiveness. Furthermore, such decrease in sampling rate allows us to increase the sampling frequency when needed with no change in number of sampled flows.

It is possible to set the position of the threshold t as one of the input parameters, or preferably the threshold can be computed dynamically by the algorithm to adapt on current input data (e.g. by using Peirce's criterion). However, we will discuss dynamic threshold computation in our future work.

Definition 8. Let $X_1, \ldots, X_k$ be primary features and let X_i be a secondary feature. We define secondary probability, which is a probability that a flow related to x through the feature X_i is selected to the sampled set, as:

$$p_s(x|X_i) = \begin{cases} d & \mathsf{c}(x|X_1, , X_k) > t \wedge RU(\mathsf{e}_{X_i}(x|X_1, , X_k)) \in I_1 \\ 1 & \text{otherwise} \end{cases}$$

$$I_1 \in [0, \varepsilon] \cup [1 - \varepsilon, 1], \quad \varepsilon \in (0, 0.5),$$

where $d \in (0, 1]$ is a parameter characterizing the decrease of incremental information value of the set with almost all identical flows ($RU \to 0$) or on the other side, mostly diverse flows ($RU \to 1$). Parameter ε determines the size of the interval, where the flows are considered as almost identical or mostly diverse.

Definition 9. Let $X_1, \ldots, X_k$ be primary features and $X_{k+1}, \ldots, X_n$ secondary features. Then the probability that the adaptive sampling will select flow x is defined as follows:

$$p(x) = p_p(x \mid X_1, \ldots, X_k) \cdot \prod_{i=k+1}^{n} p_s(x \mid X_i). \tag{1}$$

Theorem 1. Let T_{flows} be maximal number of flows selected from U into the sampled sets $\mathsf{S}_1^{(a)}, \ldots, \mathsf{S}_m^{(a)}$ by using the adaptive sampling with primary feature X_i and sampling rate s computed as:

$$s = \frac{\min\{|U|, T_{flows}\}}{\sum_{\mathsf{c}(x|X_i) \leq t} 1 + \sum_{\mathsf{c}(x|X_i) > t} \frac{\log t}{\log \mathsf{c}(x|X_i)}}. \tag{2}$$

Then it holds:

$$\overline{\mathsf{S}}^{(a)} = \lim_{m \to \infty} \left(\frac{1}{m} \cdot \sum_{i=1}^{m} |\mathsf{S}_i^{(a)}| \right) \leq T_{flows}.$$

Proof: Let us assume that $T_{flows} \leq |U|$. Then we can express T_{flows} by using sampling rate s and threshold t as follows:

$$T_{flows} = s \cdot \left(\sum_{\mathsf{c}(x|X_i) \leq t} 1 + \sum_{\mathsf{c}(x|X_i) > t} \frac{\log t}{\log \mathsf{c}(x|X_i)} \right)$$

Note that the summations sum all flows x satisfying the threshold conditions. Now we can express $\overline{\mathsf{S}}^{(a)}$ as:

$$\overline{\mathsf{S}}^{(a)} = s \cdot \left(\sum_{\mathsf{c}(x|X_i) \leq t} 1 + \sum_{\mathsf{c}(x|X_i) > t} \frac{\log t}{\log \mathsf{c}(x|X_i)} \right) - \varepsilon_p$$

where $\varepsilon_p \geq 0$ represents decrease in number of sampled flows caused by secondary probabilities. Thus computing the parameter s according to the Eq. 2 guarantees the theorem statement. Minimum assures that the sampling rate s will never be greater than 1. Analogically for the case of $T_{flows} > |U|$.

The proposed adaptive sampling is able to modify the sampling probability to reflect feature distributions of network traffic. It selects flows according to the size of their moments in order both to suppress large, visible and easily detectable events, and to relieve some interesting facts from the smaller ones, while the feature distributions are slightly shifted for the benefit of the anomaly detection.

Adaptive sampling also provides an upper bound in total number of sampled flows as stated in Theorem 1. This theorem guarantees that the total number of sampled flows does not exceed in any circumstances predefined limit.

5   Experimental Evaluation

The goal of sampling evaluation is to verify the benefits of late and adaptive sampling technique on the effectiveness of various anomaly detection methods. The verification is based on comparison of adaptive and random flow sampling techniques in early and late configuration on real network traffic data (we did not focus on packet sampling methods due to the significantly worse results when compared to flow sampling [14] as well as on sampling techniques designed for specific anomaly detection methods which reduces their usefulness and applicability). In our evaluation, we performed three types of experiments by using three different sets of data. We measured performance of each sampling method to show differences in computational complexity, then we inspected the influence of the sampling methods on traffic feature distributions and finally, we evaluated the impact of sampling on various anomaly detection methods. The reason of using different sets of data is because of the nature of the experiments: network traffic from gigabit link is suitable for measuring sampling computational performance, but it is not very feasible to label such amount of network traffic to create the ground truth for evaluating the impact on anomaly detection.

In all our experiments, we set the parameters of the adaptive sampling as follows: we set $c(x|sIP)$ as primary feature, $e_{sP}(x|sIP)$, $e_{dIP}(x|sIP)$, and $e_{dP}(x|sIP)$ as secondary features, $d = 0.8$, $\varepsilon = 0.1$ and $t = 1000$. The adaptive early sampling method computes only count statistics for primary feature and the rest of the statistics (entropies and other counts) is computed from the sampled set. The detection methods we used require computation of the following statistics: distributions of flows/bytes/packets having the same sIP/dIP (or combination of same sIP and dPrt or same dIP and sPrt), $e_{sP}(x|sIP)$, $e_{dIP}(x|sIP)$, $e_{dP}(x|sIP)$, $e_{sIP}(x|dIP)$, $e_{sP}(x|dIP)$, and $e_{dP}(x|dIP)$.

5.1   Sampling Performance

In this experiment, we evaluated performance of sampling techniques in terms of CPU time needed for sampling the input data and create statistics for anomaly detection methods. The experiment was performed with 5-minute block of network traffic from two networks (10Gb and 1Gb link) on four sampling methods: random early (Random E), adaptive early (Adaptive E), random late (Random L), and adaptive late (Adaptive L). In early sampling, the input flows are first sampled and then the system computes the statistics from the smaller sampled set, while all late sampling techniques compute statistics before sampling, which requires more computational time as you can see in Table 2. Note that the absolute values are not so important (as they depend on evaluating computational assets - we used common 4-core 3.2GHz processor) as the relative differences between the individual methods and sizes of input data.

Random E sampling method requires minimal computational time, but is devastating for any anomaly detection method as we will show further in our experimental evaluation. On the other side, both late sampling methods (Adaptive L and Random L) are most CPU time consuming, but this time is well invested w.r.t.

Table 2. CPU time in ms needed to sample two sizes of input flows by using four types of sampling techniques

Input data	Random E	Adaptive E	Random L	Adaptive L
6,200,000 → 1,000,000	9531	13182	16972	17269
700,000 → 100,000	826	1419	2386	2730

Table 3. Feature coverage of selected features by using four types of sampling methods – higher values mean better preservation

Sampling	$f^c(sIP)$	$f^c(dIP)$	$f^c(sPrt)$	$f^c(dPrt)$	$f^c(Prot)$	$f^c(Bytes)$	$f^c(Packets)$
Random E	0.023	0.016	0.070	0.067	0.406	0.071	0.080
Random L	0.023	0.016	0.070	0.067	0.391	0.071	0.080
Adaptive E	0.060	0.013	0.174	0.162	0.469	0.143	0.151
Adaptive L	0.108	0.009	0.273	0.253	0.453	0.215	0.228

consequent anomaly detection. We can see that the difference between Random E and Random L describing the extra time needed to compute all network statistics from the original network traffic is affordable considering the fact that the input data represents 5 minutes of network traffic. In Adaptive E, only count statistics for primary feature are computed before sampling in contrast with Adaptive L. Thus the difference between Adaptive L and Adaptive E expresses additional computational cost when we compute entropies from the original set instead of sampled set. Note that most detection algorithms require much more CPU time to process this amount of network traffic.

5.2 Preserving Feature Variability

In this Section, we will compare the differences of sampling techniques (random early, random late, adaptive early, and adaptive late) from the ideal sampling by using measures described in Sec. 3 to evaluate the ability of preserving feature moments. For this evaluation, we used real university network traffic (1Gb link) combined with simulated attack. The attack was motivated by the scenario, when the attacker launches a large scanning activity, which is used for hiding the more serious SSH brute force attack (but with small intensity) against other victim in the same network. The evaluated dataset of 5 minutes of network traffic has approximately 2 million flows.

The impact of above-mentioned sampling methods on feature variability of data from the second scenario is described in Table 3. You can see that Random early sampling and Random late strategy shows almost identical behavior, which is expectable because both methods do not use pre-computed statistical information for sampling procedure. As you can see, late adaptive sampling clearly outperforms the rest of the methods in almost all feature moments, which means that it preserves better the traffic feature distributions.

In our evaluation, we clearly demonstrated that the moments of the adaptive sampling are much more reversible than the moments of random sampling

Table 4. Detection effectiveness of anomaly detection methods on different types of malicious network behaviors - no sampling. Higher values are better.

No Sampling							
Behavior	Entropy	Volume	Minds	TAPS3D	Xu	XuDstIP	All
horizontal scan	0.08	0.00	0.19	0.11	0.08	0.31	0.77
Malicious	0.45	0.11	0.44	0.27	0.41	0.23	1.91
malware	0.00	0.00	0.00	0.01	0.00	0.01	0.02
p2p	0.52	0.19	0.43	0.23	0.44	0.09	1.9
skype supernode	0.00	0.00	0.00	0.00	0.00	0.09	0.09
ssh cracking	1.00	0.00	0.82	0.99	1.00	0.85	4.66
vertical scan	0.94	0.27	0.88	0.98	0.98	0.31	4.36
sum	2.99	0.57	2.76	2.59	2.91	1.89	**13.71**

technique, which makes the adaptive sampling a promising approach for preserving statistical information. However, the selection of the primary features is very crucial with respect to the detection techniques behind the sampling algorithm.

5.3 Impact on Anomaly Detection

In this Section, we will evaluate the impact of sampling strategies on the accuracy of six anomaly detection methods. Namely the method called *Minds* [8] models differences of count moments in time, *Xu* and *XuDst* [17] operate with relative uncertainty and define static classification rules, *Volume* [12] and *Entropy* [13] prediction model network traffic by using PCA, and *Taps* [16] method specially designed for scanning activities uses sequential hypothesis testing. We integrated these state of the art methods into intrusion detection device – CAMNEP [15].

To analyze the impact of sampling on above-mentioned anomaly detection methods, we used 1 day of university network traffic, which was partially classified by network security experts (around 45% of flows) into various types of network behavior. Will refer to this classification as ground truth. We computed for each anomaly detection algorithm the amount of corresponding network behavior, which is placed among 20% of the most anomalous/malicious flows and we will refer to this values as *detection effectiveness*. For example, value 1 means that all flows of an attack belongs among 20% of most anomalous flows.

The whole evaluation results are shown in Table 4 - Table 6. Table 4 describes the detection potential of each method when no sampling is performed. We will evaluate the impact of sampling by comparing all results with this baseline case. Random early and adaptive early sampling (Table 5) significantly decreases the detection capabilities described by the boldfaced right-most number, which is almost half of the original value when compared to the case where no sampling is used. On the other hand, both random late and adaptive late (Table 6) show comparable results with no sampling method. This means we can use late sampling without any significant damage to anomaly detection effectiveness.

The reason why adaptive and random sampling are similar is mainly because of the fact that the evaluated traffic does not contain larger network incidents,

Table 5. Detection effectiveness of anomaly detection methods on different types of malicious network behaviors - early sampling. Higher values are better.

Random Early Sampling							
Behavior	Entropy	Volume	Minds	TAPS3D	Xu	XuDstIP	All
horizontal scan	0.03	0.02	0.1	0.02	0.12	0.27	0.56
malicious	0.25	0.16	0.25	0.15	0.36	0.25	1.42
malware	0	0	0	0	0	0.21	0.21
p2p	0.25	0.22	0.19	0.08	0.32	0.17	1.23
skype supernode	0	0	0	0	0	0.19	0.19
ssh cracking	0	0	0	0	0	0.1	0.1
vertical scan	0.53	0.01	0.92	0.84	0.89	0.62	3.81
sum	1.06	0.41	1.46	1.09	1.69	1.81	**7.52**

Adaptive Early Sampling							
Behavior	Entropy	Volume	Minds	TAPS3D	Xu	XuDstIP	All
horizontal scan	0	0.01	0.14	0.01	0.06	0.36	0.58
malicious	0.14	0.09	0.24	0.08	0.23	0.24	1.02
malware	0	0	0	0	0	0.21	0.21
p2p	0.3	0.2	0.35	0.14	0.36	0.26	1.61
skype supernode	0	0	0	0	0	0.3	0.3
ssh cracking	0	0	0	0	0	0.07	0.07
vertical scan	0.49	0.77	0.95	0.54	0.98	0.13	3.86
sum	0.93	1.07	1.68	0.77	1.63	1.57	**7.65**

Table 6. Detection effectiveness of anomaly detection methods on different types of malicious network behaviors - late sampling. Higher values are better.

Random Late Sampling							
Behavior	Entropy	Volume	Minds	TAPS3D	Xu	XuDstIP	All
horizontal scan	0.09	0	0.18	0.06	0.11	0.31	0.75
Malicious	0.41	0.15	0.41	0.22	0.4	0.24	1.83
malware	0	0	0	0	0.02	0.23	0.25
p2p	0.49	0.22	0.46	0.2	0.45	0.19	2.01
skype supernode	0	0	0.01	0	0	0.23	0.24
ssh cracking	1	0	0.8	0	1	0.67	3.47
vertical scan	0.94	0.02	0.99	0.98	0.93	0.74	4.6
sum	2.93	0.39	2.85	1.46	2.91	2.61	**13.15**

Adaptive Late Sampling							
Behavior	Entropy	Volume	Minds	TAPS3D	Xu	XuDstIP	All
horizontal scan	0.11	0.01	0.26	0.04	0.08	0.34	0.84
Malicious	0.28	0.08	0.3	0.21	0.27	0.21	1.35
malware	0	0	0	0.01	0	0.13	0.14
p2p	0.51	0.2	0.51	0.37	0.48	0.17	2.24
skype supernode	0	0	0	0	0	0.18	0.18
ssh cracking	1	0	0.84	0.37	0.99	0.88	4.08
vertical scan	0.96	0.42	1	0.99	0.97	0.63	4.97
sum	2.86	0.71	2.91	1.99	2.79	2.54	**13.8**

which could be suppressed by the adaptive sampling. From this reason, we will return to the datasets with large scan and hidden brute force attack and show the differences of sampling methods on this specific scenario.

We divided the evaluation into two parts: first the large scan is excluded from the datasets, and second where the scan is included. The quality of detection of the brute force attack and response for each anomaly detection method is computed as follows:

$$Q = (attack - \mu)/\sigma,$$

where $attack$ is anomaly value of the attack and μ and σ is the mean and standard deviation of anomaly values of all flows. Higher positive value means that the attack is better separated from the rest of the traffic in the malicious zone. On the other hand, negative value means unsatisfactory detection result.

Figures 1(a) - (b) illustrate the results without large scan (with only small brute force attack) of every individual detection method and simple average, while Figures 1(c) - (d) are related to the second scenario (small brute force attack hidden behind extensive scans). As you can see from Figure 1(a), adaptive sampling separates the brute force attack from the rest of the traffic better than any other

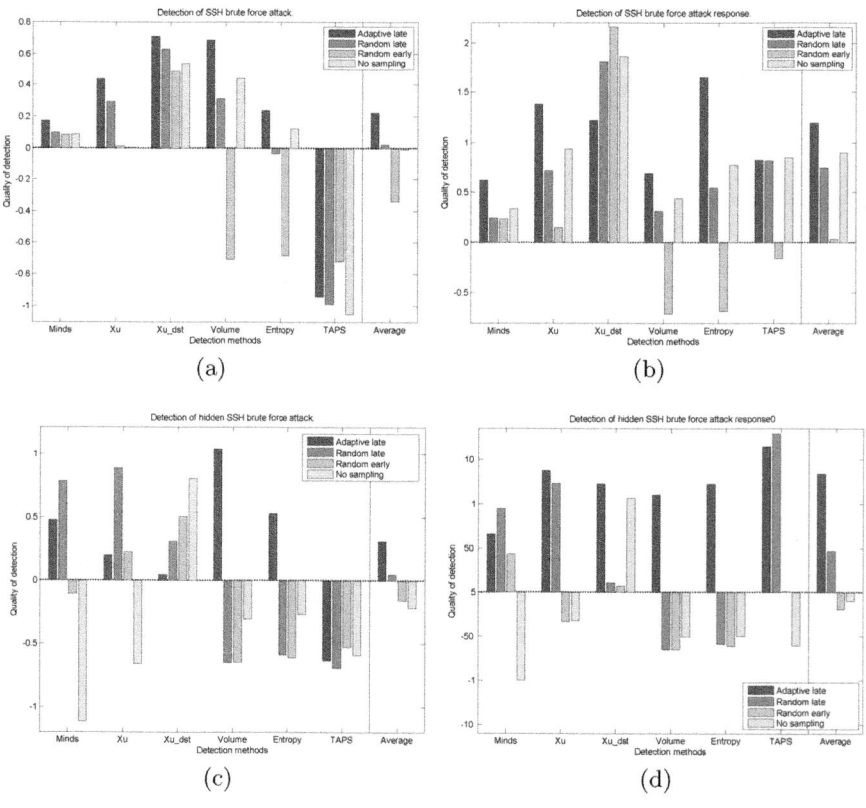

Fig. 1. Detection of quality of SSH brute force attack/response by using different anomaly detection methods and sampling techniques

technique (including when no sampling is applied), which is clearly visible in the rightmost part of the figure. On the other side, random early sampling shows very unsatisfactory results. The quality of detection of brute force response is comparable except for random early sampling, where the detection methods did not separate the attack from the traffic at all. At this point, we can say that late sampling is crucial for detecting the attack, but we did not observe great difference between random and late sampling.

The situation becomes different in the second scenario illustrated in Figures 1(c) and (d), where the adaptive late sampling is a necessity to be able to detect the attack or response. This is caused by the adaptive properties of the sampling, which decreased the amount of flows related to large scan activity allowing the detection layer to concentrate on the rest of the flows in more detail.

6   Conclusion

In this paper, we present adaptive flow-based sampling technique suitable for wide range of anomaly detection methods. This technique reduces the loss of information caused by sampling procedure with an innovative approach of suppressing large redundant entities, while emphasizing the small artifacts that may be equally important from the security perspective. To increase the sampling effectiveness, we present late sampling technique that provides exact statistical information about the original dataset, which makes this technique extremely useful for any anomaly detection method based on those statistics.

Furthermore we define formal model of the ideal sampling method and use quality metrics to evaluate various sampling properties between adaptive and random sampling algorithms. The formal model and quality metrics are general enough to allow any sampling procedure to quantify the quality of the result it provides from the anomaly detection standpoint.

According to our experiments, we demonstrated that the combination of adaptive sampling method and late sampling technique is a promising general approach that preserves well the traffic feature distributions and at the same time is able to improve the detection capabilities of the system.

Acknowledgment. This material is based upon work supported by the ITC-A of the US Army under Contract W911NF-12-1-0028 and by AFRL Rome under the Contract FA8655-10-1-3016. Any opinions, findings and conclusions or recommendations expressed in this material are those of the author(s) and do not necessarily reflect the views of the US Government. Also supported by Czech Ministry of Education grant AMVIS-AnomalyNET: MSMT ME10051.

References

1. Ali, S., Haq, I.U., Rizvi, S., Rasheed, N., Sarfraz, U., Khayam, S.A., Mirza, F.: On mitigating sampling-induced accuracy loss in traffic anomaly detection systems. SIGCOMM Comput. Commun. Rev. 40, 4–16 (2010)

2. Androulidakis, G., Chatzigiannakis, V., Papavassiliou, S.: Network anomaly detection and classification via opportunistic sampling. Netwrk. Mag. of Global Internetwkg. 23, 6–12 (2009)
3. Androulidakis, G., Papavassiliou, S.: Improving network anomaly detection via selective flow-based sampling. Communications, IET 2(3), 399–409 (2008)
4. Choi, B.-Y., Zhang, Z.-L.: Adaptive random sampling for traffic volume measurement. Telecommunication Systems 34, 71–80 (2007), doi:10.1007/s11235-006-9023-z
5. Duffield, N.: Sampling for passive internet measurement: A review. Statistical Science 19, 472–498 (2004)
6. Duffield, N., Lund, C., Thorup, M.: Properties and prediction of flow statistics from sampled packet streams. In: Proceedings of the 2nd ACM SIGCOMM Workshop on Internet Measurment, New York, NY, USA, pp. 159–171 (2002)
7. Duffield, N., Lund, C., Thorup, M.: Estimating flow distributions from sampled flow statistics. IEEE/ACM Trans. Netw. 13, 933–946 (2005)
8. Ertoz, L., Eilertson, E., Lazarevic, A., Tan, P.-N., Kumar, V., Srivastava, J., Dokas, P.: Minds - minnesota intrusion detection system. In: Next Generation Data Mining. MIT Press (2004)
9. Estan, C., Keys, K., Moore, D., Varghese, G.: Building a better netflow. SIGCOMM Comput. Commun. Rev. 34, 245–256 (2004)
10. Estan, C., Varghese, G.: New directions in traffic measurement and accounting. In: Proceedings of the 2002 Conference on Applications, Technologies, Architectures, and Protocols for Computer Communications, SIGCOMM 2002, pp. 323–336. ACM, New York (2002)
11. Hohn, N., Veitch, D.: Inverting sampled traffic. IEEE/ACM Transactions on Networking 14(1), 68–80 (2006)
12. Lakhina, A., Crovella, M., Diot, C.: Diagnosis Network-Wide Traffic Anomalies. In: ACM SIGCOMM 2004, pp. 219–230. ACM Press, New York (2004)
13. Lakhina, A., Crovella, M., Diot, C.: Mining Anomalies using Traffic Feature Distributions. In: ACM SIGCOMM, Philadelphia, PA, pp. 217–228. ACM Press, New York (2005)
14. Mai, J., Chuah, C.-N., Sridharan, A., Ye, T., Zang, H.: Is sampled data sufficient for anomaly detection? In: Proceedings of the 6th ACM SIGCOMM Conference on Internet Measurement, IMC 2006, pp. 165–176. ACM, New York (2006)
15. Rehak, M., Pechoucek, M., Grill, M., Stiborek, J., Bartos, K., Celeda, P.: Adaptive multiagent system for network traffic monitoring. IEEE Intelligent Systems 24(3), 16–25 (2009)
16. Sridharan, A., Ye, T., Bhattacharyya, S.: Connectionless port scan detection on the backbone, Phoenix, AZ, USA (2006)
17. Xu, K., Zhang, Z.-L., Bhattacharrya, S.: Reducing Unwanted Traffic in a Backbone Network. In: USENIX Workshop on Steps to Reduce Unwanted Traffic in the Internet (SRUTI), Boston, MA (July 2005)
18. Yang, L., Michailidis, G.: Sampled based estimation of network traffic flow characteristics. In: 26th IEEE International Conference on Computer Communications, INFOCOM 2007, pp. 1775–1783. IEEE (May 2007)

Detecting and Profiling TCP Connections Experiencing Abnormal Performance

Aymen Hafsaoui[1], Guillaume Urvoy-Keller[2], Matti Siekkinen[3], and Denis Collange[4]

[1] Eurecom, France
[2] Laboratoire I3S CNRS/UNS UMR 6070, France
[3] Aalto University, Finland
[4] Orange Labs, France

Abstract. We study functionally correct TCP connections – normal set-up, data transfer and tear-down – that experience lower than normal performance in terms of delay and throughput. Several factors, including packet loss or application behavior, may lead to such abnormal performance. We present a methodology to detect TCP connections with such abnormal performance from packet traces recorded at a single vantage point. Our technique decomposes a TCP transfer into periods where: (i) TCP is recovering from losses, (ii) the client or the server are thinking or preparing data, respectively, or (iii) the data is sent but at an abnormally low rate. We apply this methodology to several traces containing traffic from FTTH, ADSL, and Cellular access networks. We discover that regardless of the access technology type, packet loss dramatically degrades performance as TCP is rarely able to rely on Fast Retransmit to recover from losses. However, we also find out that the TCP timeout mechanism has been optimized in Cellular networks as compared to ADSL/FTTH technologies. Concerning loss-free periods, our technique exposes various abnormal performance, some being benign, with no impact on user, e.g., p2p or instant messaging applications, and some that are more critical, e.g., HTTPS sessions.

1 Introduction

Several access technologies are now available to the end user for accessing the Internet, e.g., ADSL, FTTH and Cellular. Those different access technologies entail different devices, e.g., smartphones equipped with dedicated OS like android. In addition, a different access technology also implies a different usage, e.g., it is unlikely that p2p applications be used as heavily on Cellular than on wired access. Even if we consider ADSL and FTTH, which are two wired technologies, some differences have been observed in terms of traffic profile [1].

Despite this variety of combinations of usage and technology, some constant factors remain in all scenarios like the continuous usage of email or the use of TCP to carry the majority of user traffic. This predominance of TCP constitutes the starting point of our study and our focus in the present work is on the performance of TCP transfers.

In this work, we aim at detecting functionally correct TCP connections – normal set-up/tear-down and actual data transfer – that experienced bad performance. The rationale behind this study is that bad performance at the TCP layer should be the symptom of

A. Pescapè, L. Salgarelli, and X. Dimitropoulos (Eds.): TMA 2012, LNCS 7189, pp. 107–120, 2012.
© Springer-Verlag Berlin Heidelberg 2012

bad performance at the application or user level. Note that this a different objective from the detection of traffic anomalies, where the focus is to detect threats against the network, e.g. DDoS [2] [3,4,5,6].

To tackle the problem, we adopt a divide and conquer approach, where we analyze separately connections that experience losses and connections that are unaffected by losses. Our analysis of connections unaffected by losses (the majority of connections) uses as a starting point a breakdown approach initially proposed in [7]. It enables to delineate, for each transfer, time periods due to the server or client thinking, or the time spent sending data. Once each connection is transformed into in a point in multidimensional space, we isolate anomalous TCP connections experiencing bad performance as those connections having high value(s) in one or several dimensions.

We apply our methodology to passive traces of traffic collected on ADSL, Cellular and FTTH access core networks managed by the same Access Service Provider. Our main contributions are as follows:

– Concerning losses, we extend to the case of multi-technology Internet access, what other studies have observed, namely that losses lead to a substantial, from 30 to 70% (median) increase of transfers times for all connection sizes (mice or elephants).
– We observed that the strategies observed on the Cellular technology to recover from losses seem more efficient than on ADSL and FTTH, as the time out durations are close to the Fast Retransmit durations.
– Concerning transfers unaffected by losses, we propose different definitions of what an abnormal performance means and exemplify the different approaches on our traces. A salient point is that our approach relies on an adequate normalization of those quantities in order to pinpoint abnormal performance independently of the actual size of the connection.
– While analyzing the connections flagged as abnormal, we relate the performance at the transport layer to the performance at the application layer. We show that in some cases, e.g., instant messaging applications, the low performance at the transport are unrelated to problems at the application layer. On the opposite, in some key client/server applications like HTTPS transfers, low performance at the transport layer might be perceived negatively by the end user.

2   Datasets

We collected packet level traces of end users traffic from a major French ISP involving different access technologies: ADSL, FTTH and Cellular. The latter corresponds to 2G and 3G/3G+ accesses as clients with 3G/3G+ subscriptions can be downgraded to 2G depending on the base station capability. ADSL and FTTH traces correspond to all the traffic of an ADSL and FTTH Point-of-Presence (PoP) respectively, while the Cellular trace is collected at a GGSN level, which is the interface between the mobile network and the Internet. Note that ADSL and FTTH clients might be behind 802.11 home networks, but we have no means of detecting it.

Table 1 summarizes the main characteristics of each trace. Each trace features enough connections to obtain meaningful statistical results.

Table 1. Traces Description

	Cellular	FTTH	ADSL
Date	2008-11-22	2008-09-30	2008-02-04
Starting Capture	13:08:27	18:00:01	14:45:02:03
Duration	01:39:01	00:37:46	00:59:59
NB Connections	1,772,683	574,295	594,169
Functionally correct cnxs	1,236,253	353,715	381,297
Volume UP(GB)	11.2	51.3	4.4
Volume DOWN(GB)	50.6	74.9	16.4

Our focus is on applications on top of TCP, which carries the vast majority of bytes in our traces. We restrict our attention to the connections that correspond to presumably valid and complete transfers from the TCP layer perspective, that we term functionally correct connections. Functionally correct connections must fulfill the following conditions: (i) A complete three-way handshake; (ii) At least one TCP data segment in each direction; (iii) The connection must finish either with a FIN or RST flag. Functionally correct connections carry between 20 and 125 GB of traffic in our traces (see Table 1). The remaining connections, which amount for almost one third of connections in our traces, consist for the vast majority of transfers with non complete three-way handshakes (presumably scans) and also a minority of connections, a few percents, for which we missed the beginning or the end of the transfer.

To have in idea of the applications present in our data sets, we performed a rough classification of traffic by identifying destination ports. It reveals that more than 84% of Cellular access connections targeted ports 80 and 443. This value falls to respectively 45% and 62% of bytes for our FTTH and ADSL traces, where we observed a prevalence of dynamic destination ports, which are likely to correspond to p2p applications.

3 On the Impact of Losses on TCP Performance

TCP implements reliability by detecting and retransmitting lost segments. The common belief is that the loss recovery mechanism of TCP is particularly penalizing for short transfers. However, several work have shown that even long transfers might be penalized by loss recovery, e.g. [7]. We confirm those results for the cases of all the access technology we consider. We further demonstrate that on the Cellular access technology, some counter measures have been put in place to limit the duration of TCP recovery phases.

3.1 Losses and Retransmission Periods

To assess the impact of TCP loss retransmission events in our traces, we use an algorithm to detect retransmitted data packets, which occur between the capture point and the remote server or between the capture point and the local (ADSL, FTTH, Cellular) client. This algorithm is similar to the one developed in [8]: we define the retransmission time as the time elapsed between the moment where we observe a decrease of the TCP sequence number and the first time where it reaches a value larger than the largest sequence number observed so far. If ever the loss happens after the observation point, we observe the initial packet and its retransmission. In this case, the retransmission time is simply the duration between those two epochs. When the packet is lost before the probe, we infer the epoch at which it should have been observed, based on the sequence numbers of packets. Note that computations of all those durations are performed at the sender side, as time series are shifted according to our RTT estimate. We (heuristically) separate actual retransmissions from network out of sequence and spurious [8] retransmission events by eliminating durations smaller than the RTT of the connection. Once losses are identified, we compute for each TCP connection, its total retransmission time.

We first report, in Table 2, on two metrics: the average loss rate and the average fraction of connections affected by loss events.

Table 2. Overall loss rates

	Cellular	FTTH	ADSL
Loss rate	4%	2%	1.2%
Fraction of connections	29%	9%	5%

We observe from Table 2 that while loss rates are quite low (our traces are too short to draw general conclusions on the loss rates in each environment), the fraction of connections affected by losses are quite high, esp. for the Cellular technology. A possible reason is that losses are due to random losses on the wireless medium, which may result in small loss episodes that affect connections irrespectively of their duration or rate.

To assess the impact of the loss recovery mechanisms of TCP, we compute the fraction of transfer time that the recovery period represents. The transfer time itself is defined as the sum of set-up (three-way handshake) and data transfer time (including loss recovery periods) for each connection. We exclude the tear-down time, where only control segments are exchanged (ACK, FIN, RST,) as this duration as no impact on application performance[1] and has been observed to be extremely long in a lot of cases - see [7]. As we further want to assess the impact of the recovery process for both short and long connections, we present results for each decile of the connection size, i.e., we report results for the 10% of smaller connections, then the next 10%, etc. Results for each access technology are presented in Figure 1. We observe that for all technologies, losses lead to a significant increase of the connection duration, between 30 and 70 % when considering the medians (bars in the center of the boxplots), irrespectively of the actual size of the transfer. We also note that the lower impact is observed for the Cellular trace, while the impact on the ADSL and FTTH traces are similar for all deciles.

We further investigate this discrepancy Cellular and ADSL/FTTH in the next section.

3.2 Delving into TCP Retransmissions

To uncover the better performance of Cellular connections observed in the last paragraph, we next distinguish between losses recovered by a retransmission time-out (RTO)

---

[1] A server might in fact be affected by long tear down times as the resources associated to the socket are unnecessarily affected to the connection.

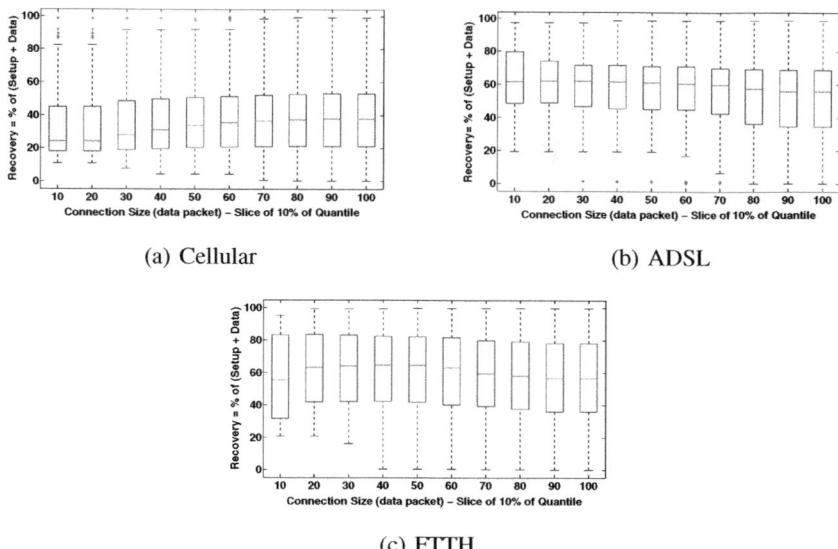

(a) Cellular

(b) ADSL

(c) FTTH

Fig. 1. Fraction of connection duration due to TCP retransmissions

and losses recovered by a fast retransmit/fast recovery (RF/R). We suppose that RTO (resp. FR/R) correspond to recovery periods with strictly less than (resp. greater or equal to) 3 duplicate acknowledgments. This definition leads to a striking result: for our traces, more than 96% of loss events are detected using RTO. This result is in line with previous studies [7].

Two factors contribute to this result. First, most of transfers are short and it is well-known that short transfers, which do not have enough in flight packets to trigger a FR/R revert to the legacy RTO mechanism. Second, long connections must often rely on RTO as the transfer, while large, consists of a series of trains (questions and answers of the application layer protocol) whose size is not large enough, in almost 50% of the cases in our traces, to trigger a FR/R.

Figure 2 plots the distribution of data retransmission time for FR/R and RTO based retransmissions. As expected, FR/R retransmission times are shorter than RTO for all access technology. However, the key result here is that under the Cellular technology, a significant attention has apparently been paid to limit the RTO duration, which results in RTO performance close to the FR/R performance. This might be due to specific mechanisms at the server side [2] or at the Access Point Name, which is the proxy that Cellular clients use to the access the Internet. Optimizing the RTO mechanism in the Cellular environment is a strategy that pays offs as the vast majority of TCP transfers rely on RTO. For instance, if we arbitrarily set a threshold in terms of abnormal performance to 1s of recovery period, we observe that with the current optimization, the fraction of abnormal recovery times is about 20% smaller in Cellular than in ADSL and FTTH scenarios.

---

[2] While the protocol stack of the end device might also play a role, most of the data packets flow from the server to the cellular client.

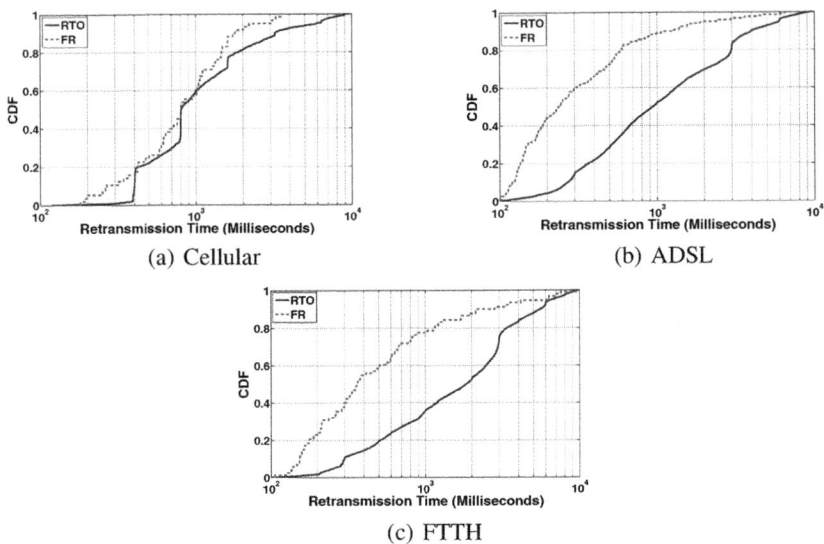

(a) Cellular

(b) ADSL

(c) FTTH

Fig. 2. Retransmission times due to FR/R or RTO

4 TCP Abnormal Performance Due to Causes Other Than Losses

4.1 Methodology

We next turn our attention to connections that are not affected by retransmissions. We are left with the set-up, data transfer and tear-down times. We did not observe long set-up times, due for instance to the loss of SYN/SYN-ACK packets. We thus do not consider set-up times in our analysis. We exclude the tear-down phase from our analysis for the same reasons as in the previous section: it does not affect client perceived performance and can bias our analysis as tear-down durations can be extremely large as compared to the actual data transfer. To highlight the above assertion, we present in Figure 3 the legacy throughput (total amount of bytes divided by total duration including tear down) and what we call the Application-Layer (AL) throughput where tear-down is excluded. We already see a major difference between those two metrics. If we are to reveal the actual performance perceived by the end user, we further have to remove the durations from the epochs where the user[3] has received all data she requested from the server (which we detect as no unacknowledged data from the server to the client in flight) and the epochs where she issues her next query. We call this metric the Effective Exchange (EE) throughput. Those three metrics (throughput, AL throughput and EE throughput) are presented in Figure 3 and we can see that they present highly different views of the achieved performance.

---

[3] The user might be a program, e.g, a mail client sending multiple mails.

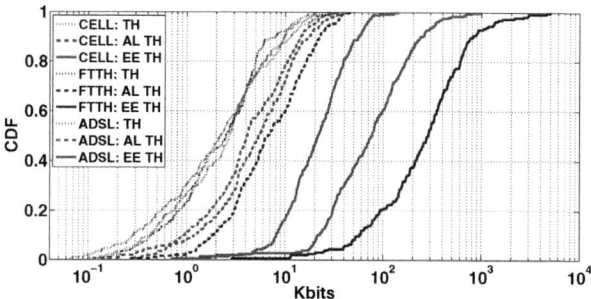

Fig. 3. Various Ways of Computing Connection Throughput

We can generalize the above approach by decomposing each transfer into 6 periods whose durations sum to the total transfer durations:

- The client[4] and server **warm-up** times, where either the client is thinking or the server is crafting data;
- The **theoretical times** computed on the client and server side, which represent the time an ideal TCP connection acting on the same path (same RTT but infinite bandwidth) would take to transfer all data from one side to the other. As a simple example, consider a TCP connection that must convey 7 data packets from a sender A to a receiver B. Assuming an infinite bandwidth, it takes $3.5 \times RTT$ to transfer the packets from A to B if we assume an initial congestion window of 1 and the use of the delayed acknowledgment mechanism.
- The difference between the transfer in one direction (say client to server) and the sum of thinking time and theoretical time is due to some phenomenon in the protocol stack, e.g. the application or the uplink/downlink capacity that slowed down the transfer. We call **pacing** this remaining duration.

Figure 4 depicts an example of our decomposition approach for the case of a browsing session.

The above methodology was presented in [7] with a different objective than detecting anomalies. We aim here at using it to isolate abnormal TCP connections. A first step is that we exclude the client side warm-ups as large thinking time at the client side should not mean anomaly. Next, we apply a normalization process on each dimension as we want to select anomalous connections irrespectively of their actual size. To do so, we apply the following normalization procedure :

- Normalized Warm-up: for each connection, we obtain the normalized warm-up as the time total warm-up divided by the number of warm-up events.
- Normalized theoretical times: for each connection, we obtain the normalized theoretical time for each direction as the total theoretical time divided by the number of packet for the corresponding direction. For long connections, the normalized theoretical times should be close to the RTT of the connection. For small connections, the speed at which the congestion window opens will constrain the normalized theoretical time.

---

[4] The client is for us, the initiator of the transfer.

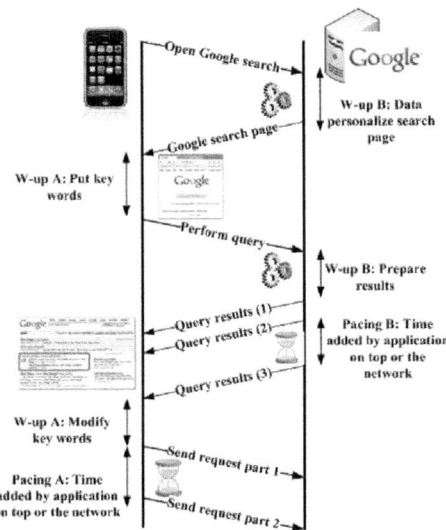

Fig. 4. Decomposition of a typical TCP transfer

- Normalized pacing: similarly to the theoretical time, we divide the total pacing time per direction by the number of packets for the corresponding direction.

We next present three different definitions of an anomaly. Note that since we excluded the Warm-up at the client side, each connection corresponds to a point in a 5 dimensional space.

- **Definition 1:** a connection is declared anomalous if its value, in any of the 5 dimension is higher than the p-th quantile for this dimension.
- **Definition 2:** a connection is declared anomalous if the sum of its values in each dimension is higher than the p-th quantile (computed over the sum).
- **Definition 3:** a connection is declared anomalous if its normalized response time, which is defined as its transfer time (set-up plus data transfer time. Again, we exclude the tear-down phase) divided by the total number of packets transferred (sum over both directions) is higher than the p-th quantile of the corresponding distribution.

Each of the above definitions has its own merits. Definition 3 is the simplest one and does not require our break-down methodology to be applied. Definitions 1 and 2 are built on our decomposition approach. Definition 1 aims at detecting outlier in at least one dimension while definition 2 aims at detecting global outliers, which might not have extremely high values in any dimension but a globally high sum.

In the present work, our objective is to understand which anomalies can be detected using our approach. We leave aside the important problem of determining which definition is the best and also, which value of p is the best. Instead, we focus on analyzing a set of connections flagged by the three definitions for an arbitrary value of p.

4.2   Selection of Anomalous Connections

We proceeded as follows to select a set of abnormal connections. Our starting point is definition 1, for which we use $p = 85$, i.e., we select a connection as an outlier if its values in any dimension is larger than the 85-th quantile in this dimension. Using this approach and this threshold value will lead to select between 15 and 75% of connections. It will be 15% if a connection that features a high value in one dimension also features a high value in all the other dimensions. Conversely, if we have disjoint sets of connections for each dimension, we obtain $5 \times 15 = 75\%$ of connections. In our case, we obtain an intermediate value of 33% of connections. Those 33% of connections correspond to 5% of the overall bytes exchanged . We next adjust the threshold in definitions 2 and 3 so as to have the same number of connections selected as in definition 1. This simply means that we set $p = 77\%$ for definitions 2 and 3. As we do not want to decide which definition is the best at this stage, we consider the intersection of the sets of connections selected using those 3 definitions. The matrix below provides the percentages of intersections using each definition:

$$\begin{pmatrix} & \text{Def. 1} & \text{Def. 2} & \text{Def. 3} \\ \text{Def. 1} & 100\% & 53\% & 55\% \\ \text{Def. 2} & 53\% & 100\% & 62.1\% \\ \text{Def. 3} & 55\% & 62.1\% & 100\% \end{pmatrix}$$

4.3   Clustering Results

The intersection of the three sets correspond to 11% of connections and 3% of bytes. We use a clustering approach to discover similarities between anomalies. We used the popular Kmeans algorithm. Using Kmeans requires both to choose (before running the algorithm) the number of clusters and also the initial choice of the centroids of the cluster. For the first problem, we rely on a visual inspection of the data using a dimensionality reduction technique (t-SNE [9]) that projects multi-dimensional data on a 2D plane, while preserving the relative distance between points. Concerning the choice of the initial centroids, we use the classical Kmeans + approach whereby 100 initial choices of clusters are considered and the best result (in terms of intra and inter-cluster distances) is picked at the end.

For our set of bad performing connections selected in the previous section, we obtained with tSNE that 4 clusters was a reasonable choice. We present the 4 clusters obtained with Kmeans in Figure 6. We use a boxplot representation for each of the five dimensions. Note that the values reported here are non-normalized values: we normalize prior to clustering but we report initial values in the boxplot representations. We also enrich the graph of each cluster with (i) the fraction of connections for each access technology and (ii) the median size of transfers (both, on top of the graphs). We further present in Figures 5(a) and 5(b) the distributions of ports per cluster and also the volumes (in bytes) per cluster, respectively.

We can observe that the size of clusters range between 17 and 38%, which means that they are relatively homogeneous in terms of size. In contrast, the clusters are quite different in terms of the applications they correspond to. Cluster 1 corresponds to HTTP/HTTPS traffic and also a significant fraction of others, where others means

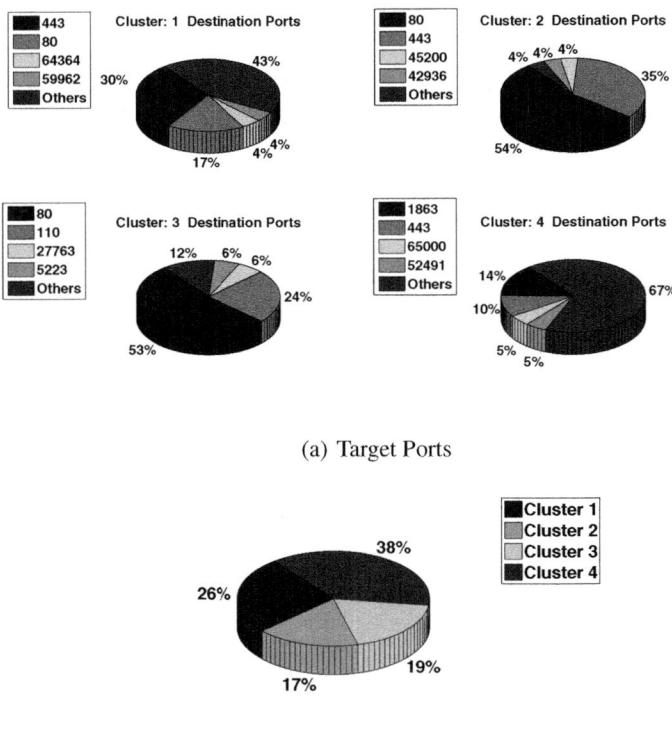

(a) Target Ports

(b) Data Volume Per Cluster

Fig. 5. Intersection

that both the source and destination ports are dynamic (a good hint of a p2p application). Cluster 2 corresponds mostly to HTTP/HTTPS traffic. Cluster 3 features mostly to HTTP and POP, while cluster 4 is dominated by dynamic ports.

When correlating the dominating ports and the fraction of connections per access technology in each cluster (on top of each plot in Figure 6), one can clearly observe that cellular access is mostly present in clusters 2 and 3 where there is little dynamic ports. In contrast, most of the ADSL and FTTH connections are in clusters 1 and 4 that contain a lot of connections corresponding to dynamic ports. Those observations are in line with intuition as dynamic ports are likely to be due to p2p applications that are more popular on ADSL/FTTH than Cellular access technology. Note that a majority of our cellular users use smartphones rather than laptop equipped with 3G dongles as we observed by mining the HTTP header (user-agent information) of their Web requests.

Let us now focus on the interpretation of the 4 clusters in Figure 6. One can adopt a quantitative or a qualitative standpoint. From a qualitative standpoint, we can observe that clusters 1 and 2 report on problems located at the server side, with extremely high warm-up or pacing times. In contrast, for clusters 3 and 4, one observes large values at both the client and the server side. If one adopts a quantitative viewpoint, the situation

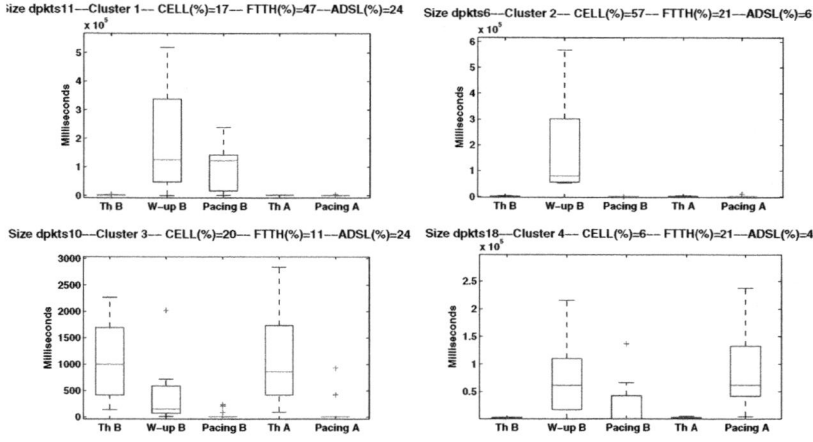

Fig. 6. Intersection: K-means Results

is quite different. Indeed, clusters 1, 2 and 4 are characterized by values (for the dimensions that characterize the anomaly) that are one to two order of magnitudes larger – in the order of tens or hundreds of milliseconds for those clusters as compared to about one second time scale for cluster 3.

Our starting point, in this work, was the hypothesis that bad performance at the TCP layer should be the symptom of a problem at the application layer or from the user point of view. A closer analysis of the clustering results reveals that while our approach indeed reveals TCP connections that perform badly, not all these connections result in bad performance from the user perspective. Let us consider the case of cluster 4. In this cluster, one observes 14% of connections using port 1863, which is the Microsoft messenger port. It is clear that here, the bad performance at the transport layer is due to the fact that the client and the server are two humans that (normally) think before typing and type relatively slowly. Bad performance at the transport layer is thus unrelated to bad performance at the application layers. Also, still for cluster 4, one observes a large fraction of dynamic ports, which is likely to be due to p2p applications. It is known that p2p applications tend to throttle the bandwidth offered to other peers. This is why we observe limitations at both the client and server side, which are the two ends of the application. It is difficult to categorize the bad performance of p2p applications as an anomaly from the user perspective since users are in general patient when it comes to download content since they treat this traffic as a background traffic (that should not interfere with their current interactive traffic, typically browsing ; hence the rate limiters in p2p applications).

The situation is different for clusters 1 and 2, for which anomalies are clearly located at the server side and a significant share of the traffic is due to HTTP and HTTPS. We observed typically large values for HTTPS connections that presumably (given the IP address) correspond to electronic payment transactions. We leave for future work a more in depth analysis of the servers flagged as anomalous (e.g., which fraction of their traffic is indeed anomalous, do they serve only a specific type of clients, e.g. cellular clients) and we present in the next section, some typical examples of anomalies that we observed, related to warm-up or pacing problems, in order to make our case more concrete.

4.4   Examples of Anomalies

Large Warm-up B. We report in Figure 7 an example of large warm-up time at the server side,observed by a client behind an ADSL access. We notice that the acknowledgment received from the server indicates that the query (GET request) has been correctly received by the server, but it takes about 4.5 seconds before the client receives the requested object (a png image in this case).

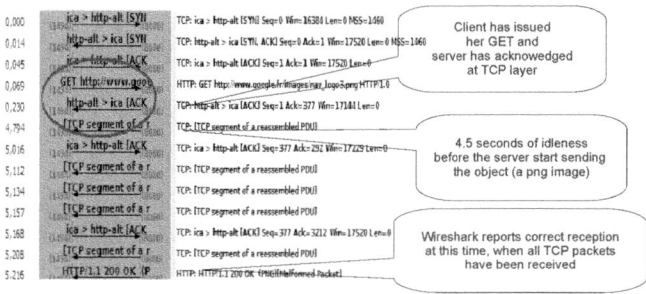

Fig. 7. Abnormal Long Response Time at The Server Side (Warm-up B value)

Large Pacing B. We report in Figure 8 an example of large pacing time for a Gmail server, observed by a client behind an FTTH access. We notice that the acknowledgment received from the server indicates that the **GET request** has been correctly received by the server. The server sends data until the last TCP segment which is delayed by 27.6 seconds, before the client receives the object.

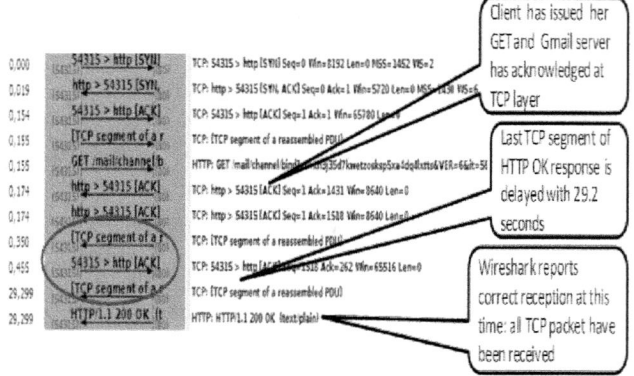

Fig. 8. Webmail: Large Pacing B in Gmail Server

Large Pacing A. We report in Figure 9 an example of large Pacing A in the case of an HTTPS connection for a Cellular client. After a successful three way handshake, the user authenticates and exchanges data with the server. If we focus on the client last data packet, we observe that it is delayed by more than 200 seconds compared to the previous data packet. This introduces a large idle time in the transfer. We see through this example that the application can play an important role in data scheduling at the network layer, which can have a detrimental impact in terms of perceived performance.

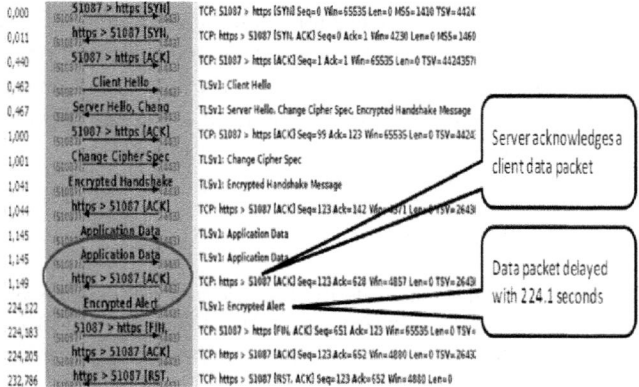

Fig. 9. Large Pacing A

5   Conclusion

In this paper, we have shed light on the problem of detecting and understanding TCP connections that experience low performance. We jointly analyzed network traces collected on a variety of networks that reflect the way people are accessing the Internet nowadays.

Losses in the network, while rare, significantly deteriorate performance. This result is not new but the added value of our work is to show that any transfer that experiences losses suffers, irrespectively of the access technology or its exact size. Furthermore, our approach of jointly profiling different access technologies enabled us to highlight that more attention is paid to limiting the impact of losses in cellular than on ADSL/FTTH networks apparently. We relate this discrepancy to the shorter durations of time out durations on the cellular network. While those results need to be confirmed over longer traces, the extent to the difference (a factor of 2 in the fraction of time required to recover losses on Cellular trace as compared to our ADSL/FTTH traces) suggests it can not be a mere coincidence. As future work, we intend to investigate whether optimizations of the recovery mechanism of TCP could be proposed based on the observations we made.

For the majority of connections that do not experience losses, we propose several approaches to detect outliers. Each of them accounts for the size of the connections so that not only long connections be flagged as experiencing abnormal performance. We exemplify the different approaches by analyzing their intersection set, i.e., the set of connections flagged as abnormal, whatever the definition is. We use a clustering approach to form groups of similar abnormal connections. We enrich those groups with additional information like the distribution of ports per group, to understand whether low performance at the transport layer is a symptom of bad performance at the application layer. It turns out that the relation between the transport and the application layer is complex. There are cases, e.g., instant messaging, where the two are fully unrelated. This is also partly the case with p2p transfers as users are resilient to low performance at the transport layer as long as they eventually obtain the content they want. On the other hand, we observed low performance for a significant number of cases, e.g. HTTP and HTTPS transfers where the user might consider that the application is misbehaving. As future work, we intend to profile more precisely the latter set of connections (where bad performance at the transport and application layer are apparently related) to better understand the extent of those anomalies.

References

1. Vu-Brugier, G.: Analysis of the impact of early fiber access deployment on residential internet traffic. In: 21st International Teletraffic Congress, ITC 21 2009, pp. 1–8 (September 2009)
2. Barford, P., Kline, J., Plonka, D., Ron, A.: A signal analysis of network traffic anomalies. In: Proceedings of the 2nd ACM SIGCOMM Workshop on Internet Measurment, IMW 2002, pp. 71–82. ACM, New York (2002)
3. Lakhina, A., Crovella, M., Diot, C.: Diagnosing network-wide traffic anomalies. In: ACM SIGCOMM, pp. 219–230 (2004)
4. Soule, A., Salamatian, K., Taft, N.: Combining filtering and statistical methods for anomaly detection. In: Proceedings of the 5th ACM SIGCOMM Conference on Internet Measurement, IMC 2005, p. 31. USENIX Association, Berkeley (2005)
5. Tellenbach, B., Burkhart, M., Sornette, D., Maillart, T.: Beyond Shannon: Characterizing Internet Traffic with Generalized Entropy Metrics. In: Moon, S.B., Teixeira, R., Uhlig, S. (eds.) PAM 2009. LNCS, vol. 5448, pp. 239–248. Springer, Heidelberg (2009)
6. Silveira, F., Diot, C., Taft, N., Govindan, R.: Astute: detecting a different class of traffic anomalies. SIGCOMM Comput. Commun. Rev. 40, 267–278 (2010)
7. Hafsaoui, A., Collange, D., Urvoy-Keller, G.: Revisiting the Performance of Short TCP Transfers. In: Fratta, L., Schulzrinne, H., Takahashi, Y., Spaniol, O. (eds.) NETWORKING 2009. LNCS, vol. 5550, pp. 260–273. Springer, Heidelberg (2009)
8. Jaiswal, S., Iannaccone, G., Diot, C., Kurose, J., Towsley, D.: Measurement and classification of out-of-sequence packets in a tier-1 ip backbone. IEEE/ACM Trans. 15(1), 54–66 (2007)
9. van der Maaten, L.J.P.: t-distributed stochastic neighbor embedding, http://homepage.tudelft.nl/19j49/t-SNE.html

Geographical Internet PoP Level Maps

Yuval Shavitt and Noa Zilberman

School of Electrical Engineering, Tel-Aviv University, Israel

Abstract. We introduce DIMES's geographical Internet PoP-level connectivity maps, created using a structural approach to automatically generate at world scale. We provide preliminary results of the algorithm and discuss the properties of the generated maps as well as their global spread.

1 Introduction

The topology of the internet and specifically the connectivity between its nodes on different levels of aggregation are an important field of research. Internet maps are presented in several levels of aggregation: from the AS level, which is the most coarse, to the finest level of routers. An interim level of aggregation is presented by PoP level maps of the Internet.

A Point of Presence (PoP) is a location where service providers place multiple routers. The equipment placed in a PoP is used to serve a certain area and to connect to other PoPs within the AS or in other ASes. Thus, for studying the Internet evolution and for many other tasks, PoP maps give a better level of aggregation than router level maps with a minimal loss of information. PoP level graphs can be used to examine the size of each AS network by the number of physical co-locations and their connectivity instead of by the number of its routers and IP links, which is an important contribution. The PoPs are annotated with a geographical location and PoP size. Using PoP level graphs one can detect important nodes in the network, understand network dynamics, examine types of relationships between service providers as well as routing policies and more.

While aggregating IPs to AS-level is a fairly simple task, PoP level maps are more difficult to create. Andersen *et al.* [1] used BGP messages for clustering IPs and validated their PoP extraction based on DNS. Rocketfuel's [7] generated PoP maps using tracers and DNS names. The iPlane project also generates PoP level maps and their connectivity [4] by first clustering IP interfaces into routers by resolving aliases, and then clustering routers into PoPs by probing each router from a large number of vantage points and using the TTL values to estimate the length of the reverse path, with the assumption that reverse path length of routers in the same PoP will be similar.

This paper presents PoP level connectivity maps generation and analysis, based on an algorithm described in [3].

A. Pescapè, L. Salgarelli, and X. Dimitropoulos (Eds.): TMA 2012, LNCS 7189, pp. 121–124, 2012.
© Springer-Verlag Berlin Heidelberg 2012

2   PoP Level Maps Construction

A PoP is a group of routers which belong to a single AS and are physically located at the same building or campus. The algorithm we use for PoP extraction looks for bi-partite subgraphs with delay constraints in the IP interface graph of an AS; no aliasing to routers is needed [3]. The bi-partites serve as cores of the PoPs and are extended with other near by interfaces.

To identify the geographical location of a PoP, we use the geographic location of each of its IPs. As all the PoP IP addresses should be located within the same campus, the location confidence of a PoP is significantly higher than the confidence that can be gained from locating each of its IP addresses separately. The location of an IP address is obtained from numerous geolocation databases, and the PoP's location is set to the median of all PoP's IP locations. Every PoP location is assigned a range of convergence, representing the expected location error range based on the information received from the geolocation databases. A detailed discussion of the extraction and geolocation algorithms is provided in our previous works [3,6].

Directed PoP level links are aggregated from their corresponding directed IP level edges. We keep the statistics of the IP level edges to infer the reliability of the link delay estimation.

The collected dataset for PoP level maps is taken from DIMES [5]. We use all traceroute measurements taken during weeks 42 and 43 of 2010, totaling 33 million, which is an average of 2.35 million measurements a day. The measurements were collected from over 1308 vantage points, which are located in 49 countries around the world.

The 33 million measurements produced 9.1 million distinct IP level edges (no IP level aliasing was performed). Out of these, $258K$ edges had less than the median delay threshold, and had sufficient number of measurements to be considered by the PoP extraction algorithm. A total of 4098 PoPs where discovered, containing 67422 IP addresses. Although the number of discovered PoPs is not large, as the algorithm currently tends to discover mainly large PoPs while missing many access PoPs, the large number of IP addresses and the spread around the world allow a large scale and meaningful PoP level connectivity evaluation.

3   PoP Level Maps Analysis

The PoP level connectivity map generated from the data set contains 86760 links, which are an aggregation of 1.65 million edges. Out of the 4098 discovered PoPs in week 42, 2010, 4091 have at least one PoP level link. 2405 PoPs have outgoing links, and 4073 PoPs have incoming links. Out of those, 18 PoPs have only outgoing links and 1686 have only incoming links. Note that a PoP without any PoP level links, or a PoP with only incoming or outgoing links still have additional IP-level connecting edges. As the full map is too detailed to display, a partial map is shown in Figure 1, demonstrating the connectivity between 430 ASes on PoP level.

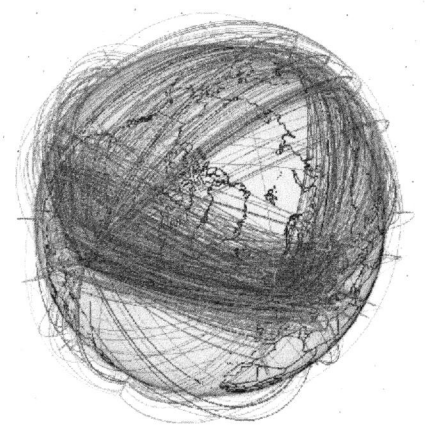

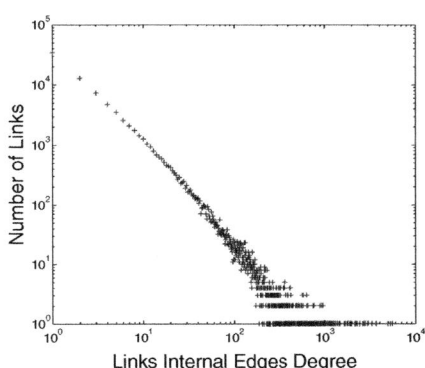

Fig. 1. An Internet PoP Level Connectivity Map - A Partial Map of Week 42, 2010

Fig. 2. Number of Edges within a Link vs. Number of PoP Level Links

Almost all the IP edges that are aggregated into links are unidirectional: 99.2%. This is a characteristic of active measurements: the vantage points are limited in number and locations, thus most of the edges can be measured only one way. However, at PoP links level, 6.5% of the links are bi-directional: eight times more than the bi-directional edges. This demonstrates one of the PoPs strengths, as it provides a more comprehensive view of the networks' connectivity without additional resources. The average number of edges within a unidirectional link is 7.5, and the average number of edges within a bidirectional link is 44.7. This is not surprising, as it is likely that most of the bidirectional links will connect major PoPs, within the Internet's core and thus be easily detected.

The number of links per PoP is shown in Figure 3. The figure shows the total number of links per PoP, the number of outgoing links (source PoP) and the number of incoming links (Destination PoP). Looking at the number of links by source PoP, 42.4% of the PoPs have only one or two links to other PoPs and 70% of the PoPs are connected by 10 links or less. Many of the links are between PoPs that are co-located, which we define as links with a minimal delay of 1mS or less, and 62.7% of the PoPs have such links. Looking at the number of links by destination PoP, 46% of the PoPs are connected by 10 links or less and the average number of links per PoP is 21. Most of the PoPs are connected to PoPs outside their AS: 71.5% of the source PoPs and 99% of the destination PoPs.

Figure 4 shows the minimum, weighted average, and maximal delay per link, plotted on a log-log scale with the delay (X-axis) measured in milliseconds. The solid black line shows the cumulative number of measurements up to a given link delay. We omit from this plot links that include only a single edge, which distort the picture as their minimal, weighted average, and maximal delay are identical. An interesting attribute of this plot is that all three plotted delay parameters behave similarly and are closely grouped. As all the links are an aggregation of multiple edges, this indicates the similarity in the delay measured on different edges.

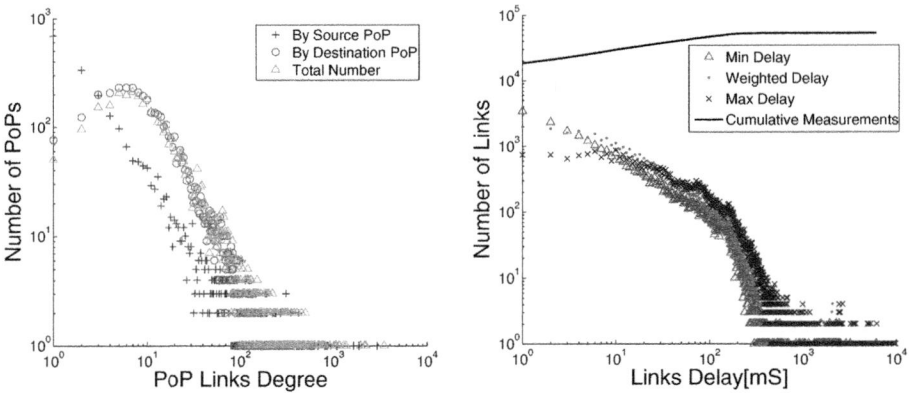

Fig. 3. Number of Links per PoP vs. Number of PoPs

Fig. 4. Links Delay vs. Number of Links

One can also see that most of the measurements represent a delay of 200ms or less, and that the extreme cases are rare (see the cumulative measurement line).

4   Conclusion and Future Work

In this paper we presented DIMES's PoP-level connectivity maps. The PoP level connectivity maps provide a new look at the Internet's topology with a better level of aggregation than router level maps and more information than AS level maps. We presented preliminary analysis of the generated maps and discussed some of its properties. We believe that these maps can be used to improve noise cleaning from measurements. The PoP level links maps are now available through the DIMES website [2] for download, and can be helpful to researchers in the fields of complex networks, Internet topology, Geolocation, and more.

References

1. Andersen, D.G., Feamster, N., Bauer, S., Balakrishnan, H.: Topology inference from BGP routing dynamics. In: IMC, pp. 243–248 (2002)
2. DIMES, http://www.netdimes.org/
3. Feldman, D., Shavitt, Y., Zilberman, N.: A structural approach for PoP geolocation. Computer Networks (2012)
4. Madhyastha, H.V., Anderson, T., Krishnamurthy, A., Spring, N., Venkataramani, A.: A structural approach to latency prediction. In: IMC 2006 (2006)
5. Shavitt, Y., Shir, E.: DIMES: Let the Internet measure itself. ACM SIGCOMM Computer Communication Review 35 (October 2005)
6. Shavitt, Y., Zilberman, N.: A geolocation databases study. IEEE Journal on Selected Areas in Communications 29(9) (December 2011)
7. Spring, N., Mahajan, R., Wetherall, D.: Measuring ISP topologies with Rocketfuel. In: ACM SIGCOMM, pp. 133–145 (2002)

Distributed Troubleshooting of Web Sessions Using Clustering

Heng Cui and Ernst Biersack

EURECOM, Sophia Antipolis, France
firstname.lastname@eurecom.fr

Abstract. Web browsing is a very common way of using the Internet to, among others, read news, do on-line shopping, or search for user generated content such as YouTube or Dailymotion. Traditional evaluations of web surfing focus on objectively measured Quality of Service (QoS) metrics such as loss rate or round-trip times; In this paper, we propose to use K-means clustering to share knowledge about the performance of the same web page experienced by different clients. Such technique allows to discover and explain the performance differences among users and identify the root causes for poor performances.

Keywords: Web Browsing, Home Networks Measurement, Troubleshooting.

1 Introduction

Web browsing is a very common way of using the Internet access as it allows to access to a wealth of information. Since there is a "human in the loop", the time it takes to render a Web page should be small in order to assure a good user experience.

Traditional evaluations of the web surfing mainly use Quality of Service (QoS) metrics that are easy to measure such as packet loss rate or round trip times. In this paper, we propose a methodology which combines a browser plugin and lower level capturing to evaluate web page performances, and we use clustering to share the performance metrics. We demonstrate that this clustering is suitable to compare and explain the experiences among different clients and further identify root causes for poor performance.

This work builds on our previous work [1] where we presented the measurement architecture but did not carry out any systematic analysis of the measurements. An extended version of this work is available in our technical report [2].

Different methodologies of troubleshooting user's network connections are proposed in the literature. Joumblatt et al. [4] propose HostView, an end-system to collect user generated packets and store all the information into a centralized server. Netalyzr [5] is a web-based diagnostic tool that to analyze and debug end-user's connectivity properties such as bandwidth, proxy, etc. Compared to these works, our work focuses only on Web browsing.

A. Pescapè, L. Salgarelli, and X. Dimitropoulos (Eds.): TMA 2012, LNCS 7189, pp. 125–128, 2012.

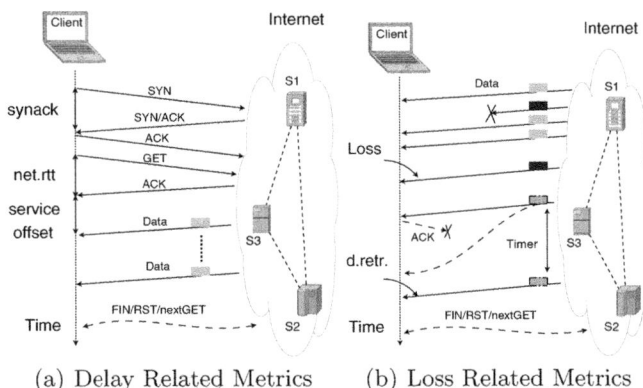

(a) Delay Related Metrics   (b) Loss Related Metrics

Fig. 1. Metrics Related to Download One Element In a Web Page

2   Methodology

We set up an experimental platform that uses a firefox plugin and normal packet capture to measure both, page level information such as full page load time, and lower packet level metrics such as RTT or loss rate. The firefox plugin, among others, binds each HTTP query initiated by the browser to its associated firefox window/tab and measures the full load time for that Web page. We use packet capture (libpcap format) to obtain raw packet traces that are loaded into a database for post-processing. For details about the architecture and how we combine the measured records, we refer the readers to our previous work [1].

A typical web page can contain up to hundreds of elements. To fully render the Web page, the browser needs to load all these elements. The typical procedure for loading one element is shown in Fig.1. As is shown in Fig.1(a), for download of each element, we extract metrics as shown in Fig.1 from the packet trace. For detailed description of our defined metrics, please refer to our extended report [2]. To describe the performance of a Web *session*, we use the metrics computed for each Web *element* and compute the **mean** over all the values of the given metric to obtain a **Key Performance Index(KPI)**, which consists of

$$[nr., SYNACK, NET.RTT, SERV.OFF., RTT, LOSS, D.RETR.]$$

Note that $nr.$ is the number of distinct GET requests during a complete web session, which provides an estimation of the size of the Web page in terms of the number of elements. Meanwhile, for the KPI, we focus on metrics captured by TCP connections, and currently we ignore DNS queries since TCP connections already provide rich enough information; and DNS pre-fetching is widely supported by recent browsers and this causes DNS queries to occur before a real web page surfing, which makes the lookup time less useful. However, we plan to study the effects of DNS (e.g. response time, response IP, TTL, etc.) on the web browsing experiences in the future. To compare results among different homes, we use clustering: we first normalize all the measured KPI metrics

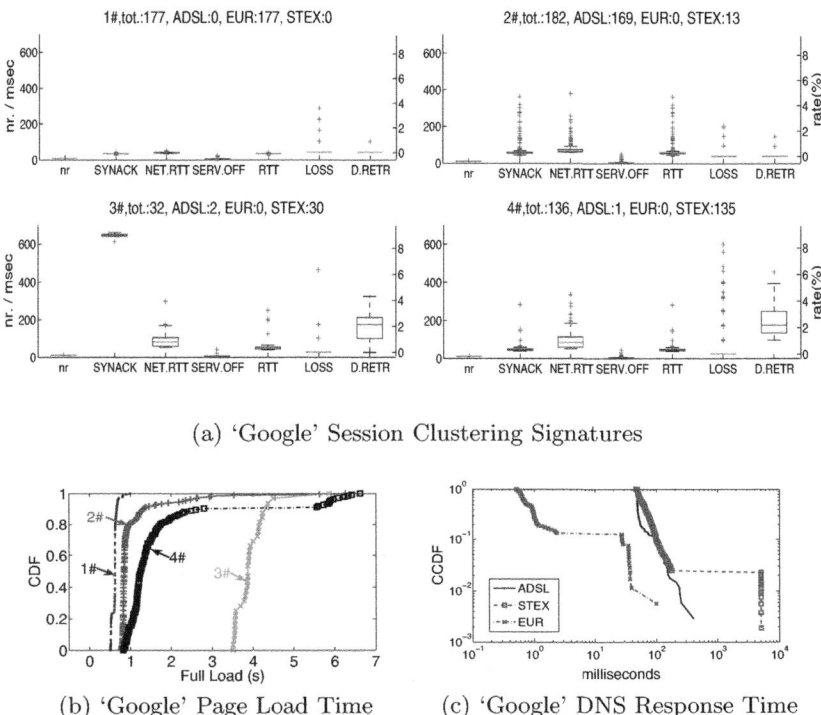

(a) 'Google' Session Clustering Signatures

(b) 'Google' Page Load Time   (c) 'Google' DNS Response Time

Fig. 2. 'Google' Sessions in Three Homes

into the range [0,1], and then use the well known *kmeans* algorithm, which is an un-supervised classification algorithm that does not need any training. The tricky point in *kmeans* is how to set a-priori the number of clusters. We use *four clusters throughout the paper,* and refer for a more detailed discussion to our technical report [2]. Furthermore, we run the *kmeans* ten times and keep the one with smallest distance error.

3 Troubleshooting Case Study

In this section, we show a simple example of how to use clustering to troubleshoot web sessions. The main idea behind this is to share browsing knowledge among different clients located in different homes. We emulate users browsing (i) at home connected via an ADSL connection, (ii) in the office at Eurecom connected via a 100 Mb/s link to the Internet and (iii) in a student residence. All the machines are located in France. We refer to the ADSL home, Eurecom office Ethernet and the student residence connections as **'ADSL'**, **'EUR'**, and **'STEX'** respectively. The experiments are done during the same evening in three homes and last for around 8 hours each. Both, the 'EUR' and 'STEX' client computers are connected via a wired connection, while the 'ADSL' client computer is physically very close to the Access Point and uses a wireless connection. We clear the browser cache at the end of each Web session.

As an illustration of how the comparison of the performance of the access to *the same Web page across different clients* helps identify the influence of problems specific to a client, we use the `Google` Web page, and show the results in Fig.2. We know that Google works very hard to keep the page download times as low as possible by placing servers close to the clients and also by keeping the number of elements of its Web page low. We see that the requests from 'EUR' are grouped in cluster 1 and from 'ADSL' are grouped in cluster 2. On the other hand, the requests in clusters 3 and 4 are issued almost exclusively from the 'STEX' client. The fact that the 'STEX' client experiences higher delays and also loss can be seen in its KPIs. Cluster 3 is interesting because of its large SYNACK values; it turns out that for cluster 3 on average one out of five SYN requests to establish a TCP connection does not get answered and must be retransmitted after timeout. Since the retransmission timeout for the SYN packet is three seconds, we get $SYNACK = \frac{(50+50+50+50+3050)}{5} = 650(ms)$. The long tail for the page load times in cluster 4 is due to the long DNS response time for some of the sessions. Since the 'STEX' client experiences more packet loss than the other two clients. We can clearly identify in Fig.2(c) that around 3% of the DNS queries need to be retransmitted.

4   Future Work

As future work, we plan to deploy our system on more end-users and also check how the number of clusters for kmeans is affected by the number of users. We also plan to extend our system to work in real time and in a distributed fashion by making the agents in the different locations communicate and perform distributed clustering [3].

References

1. Cui, H., Biersack, E.: Trouble Shooting Interactive Web Sessions in a Home Environment. In: ACM SIGCOMM Workshop on Home Networks, Toronto, Canada (August 2011)
2. Cui, H., Biersack, E.: On the Relationship Between QoS and QoE for Web Sessions. Technical Report RR-12-263, EURECOM, Sophia Antipolis, France (January 2012), http://www.eurecom.fr/~cui/techrep/TechRep12263.pdf
3. Datta, S., Giannella, C., Kargupta, H.: Approximate Distributed K-Means Clustering over a Peer-to-Peer Network. IEEE Transactions on Knowledge and Data Engineering 21, 1372–1388 (2009)
4. Joumblatt, D., Teixeira, R., Chandrashekar, J., Taft, N.: HostView: Annotating End-host Performance Measurements with User Feedback. In: ACM Sigmetrics Workshop, HotMetrics, New York, NY, USA (June 2010)
5. Kreibich, C., Weaver, N., Nechaev, B., Paxson, V.: Netalyzr: Illuminating The Edge Network. In: IMC 2010: Proceedings of the 10th Annual Conference on Internet Measurement, pp. 246–259. ACM, New York (2010)

Using Metadata to Improve Experiment Reliability in Shared Environments

Pehr Söderman, Markus Hidell, and Peter Sjödin

TSLab, School of ICT
KTH Royal Institute of Technology
SE-100 44 Stockholm, Sweden
{pehrs,mahidell,psj}@kth.se
http://www.tslab.ssvl.kth.se/

Abstract. Experimental network research is subject to challenges since the experiment outcomes can be influenced by undesired effects from other activities in the network. In shared experiment networks, control over resources is often limited and QoS guarantees might not be available. When the network conditions vary during a series of experiment unwanted artifacts can be introduced in the experimental results, reducing the reliability of the experiments. We propose a novel, systematic, methodology where network conditions are monitored during the experiments and information about the network is collected. This information, known as metadata, is analyzed statistically to identify periods during the experiments when the network conditions have been similar. Data points collected during these periods are valid for comparison. Our hypothesis is that this methodology can make experiments more reliable. We present a proof-of-concept implementation of our method, deployed in the FEDERICA and PlanetLab networks.

Keywords: Clustering, FEDERICA, Measuring, Metadata, PlanetLab.

1 Introduction

Experimental evaluation of protocols in large-scale networks is an important methodology in networking research. However, researchers rarely have access to large, dedicated infrastructures where experiments can be performed without being influenced by other activities. Instead, researchers are faced with the challenge of performing experiments on shared networks where there are other activities going on that could influence the outcome of the experiments: there can be other traffic that takes up capacity in routers, switches and links, as well as contention for CPU and memory resources in computing nodes, management activities that influence the network, and so on. If the researcher wishes to avoid this, the main alternatives are to do the experiments in an isolated laboratory environment, to use an emulated network environment, or to resort to simulations. Even though these methods are highly valuable tools for networking research, they cannot substitute experimentation in large scale networks.

A. Pescapè, L. Salgarelli, and X. Dimitropoulos (Eds.): TMA 2012, LNCS 7189, pp. 129–142, 2012.

The goal of this work is to explore tools that aid in running experiments on shared networks such as PlanetLab [18] and FEDERICA [26]. The tools are primarily intended for experimental work that involves running a series of experiments over a period of time, and then comparing the results. This could for example be a study that aims to measure throughput as a function of message size for a new protocol, or an investigation of how available bandwidth and latency to certain Internet nodes change over time. This means that several experiments will be performed, possibly with different configurations. In order to be able to draw conclusions from the results, the researcher needs to be able to compare results between different experiments. However, if the conditions on the network vary too much, different experiments may not be possible to compare. So then the question is: what experiments are performed under similar conditions, so that the data from those experiments can be compared?

1.1   Related Work

A fundamental approach to gain confidence in of experiment results is to gather additional information about the experiment environment. This information, known as *supplementary information* or *metadata*, can be used to identify and reduce errors in an experiment using statistical methods [14]. While the value of metadata is well known [17], relatively little attention has been devoted to its use in shared experiment networks.

Roscoe [20] observes that while it is hard to reproduce experiment conditions in PlanetLab, it might be possible, through continuous measurements of the network, to sufficiently *characterize* the network to allow a comparison of experiments. Roscoe calls this a *weather service*. Albrecht [1] concludes that an approach based on gathering information about the experiment environment and then later reproducing the experiments by restoring the environment is currently not feasible in PlanetLab, since the tools that would be required for this are currently missing.

With this in mind the researchers use other methods to reduce the impact of external factors on experiments when working in networks such as PlanetLab or FEDERICA. Two main strategies can be identified to reduce the impact of other network activities on the experiment, *scheduling* and *data cleaning*.

Scheduling involves careful planning to avoid running experiments in a way that competes with other activities. Experiments are scheduled to run during periods of low load, for instance, or on parts of the network that are underutilized. Spring et al. [25] propose the use of resource reservation and taking node load into consideration when deploying experiments in shared environments. Yet another approach is to randomly schedule experiments in order to even out the effects of other activities.

An example of such strategies can be found in the performance measurements of peer-to-peer streaming protocols by Seibert et al. [22]. In the experiments, nodes and execution times are randomly chosen, but only nodes with high available bandwidth and low latency are used. The wide range of nodes reduces the risk that a single node impacts the experiment results too much, and at the same time the high bandwidth ensures that the experiment is not limited by local bandwidth.

Data cleaning is the identification and exclusion of data points in order to improve the quality of measurement data [19], for instance in order to reduce bias and variance. It is also possible to "sanitize" the measurement data after a completed experiment by excluding data points that deviate too much from the expected, assuming that the deviations are caused by disturbances. This is known as "outlier elimination". An example of data cleaning can be found in [10] where nodes are discarded from the experiment if the measurement process does not start reliably. It is also possible to apply event detection algorithms [7] on metadata and thereby try to identify points or periods when significant events might have impacted the experiment.

1.2   Contributions

Our primary contribution is a novel, systematic method where we collect metadata in the background while the experiments are running. We use the metadata to identify periods when network conditions are similar, and use this as a basis for dividing experiments into groups in such a way that experiments within a group are performed at similar conditions, and hence suitable for comparison. We present a *clustering*[1] algorithm for identifying periods of similar network conditions from the metadata, along with an investigation of suitable parameter values for the clustering process.

To evaluate our method we run an experiment in the FEDERICA network. In parallel with the experiment, we perform a set of background measurements, which are used for metadata. During the experiment we also introduce disturbances in the network. We then verify that we can use our method to analyze the metadata and filter out the periods with disturbances. Furthermore, we investigate how the quality of the experiment data is improved if we only use data from periods of comparable network conditions, in contrast to using non-filtered experiment data.

Moreover, we use our method on measurement data from PlanetLab, in order to investigate the time span over which PlanetLab nodes provide comparable experiment conditions.

1.3   Outline

The rest of the paper is organized as follows. In Section 2 we give a detailed overview of the approach we have taken to the problem, and describe the algorithm we have used. Section 3 is an evaluation of our method in FEDERICA using a sample experiment, and in Section 4 we continue the evaluation using monitoring data from PlanetLab. Finally, Section 5 concludes the paper, and discusses experiences and possible further work.

---

[1] Clustering is a generic term for unsupervised algorithms that split data into subsets by minimizing a measure of similarity between two sets of observations.

2 Approach

The basis of our approach is to perform measurements in the background while the experiments are running. The results from these background measurements are used as metadata. Each data point in the metadata is a combination of a timestamp and one or several measurement values. In case the collection of metadata takes a significant time we also include a time interval for the collection. The metadata is analyzed in order to identify periods that have similar network conditions.

The data points from the experiments can then be grouped according to network conditions so that data points gathered during similar network conditions are grouped together. The idea is that data points that are in the same group are valid for comparison. We believe this helps to improve experiment validity, since it increases the confidence that comparisons of data are valid.

It is essential that the grouping of experiment data is performed by only examining the metadata, and not the experiment data itself. To do this, we first divide the metadata into time periods. We thereafter use statistical methods to compare and bundle together time periods into sets of time periods. The decisions of which time periods can be bundled together is made according to a *distance measurement function*. This procedure is a form of *clustering* of metadata values. The goal is that experiment data points gathered at time periods belonging to the same set should be suitable for comparison.

To summarize, the approach has three main steps:

1. A series of experiments is run, gathering sample data. In parallel with the experiment, we collect information about the experiment environment—the metadata. Experiment data and metadata are timestamped.
2. The clustering algorithm is performed on the metadata, resulting in a number of metadata cluster where each cluster represents periods of similar network conditions.
3. Once periods of similar conditions have been identified, we examine the actual experiment data. The experiment data is grouped according to the metadata clustering—experiments that run at periods of similar network conditions are grouped together.

2.1 Clustering Algorithm

So far, we have not discussed at any level of detail what it means that periods have similar network conditions. This is something that we will leave open, to some extent, since it is part of the experiment design —it depends on the character of the experiments, the network where the experiments run, and so on. However, for the next step we need a more precise measure of similarity between metadata measurements over time. This measure is part of the design of the clustering algorithm.

A number of algorithms can be considered for analyzing metadata to identify periods of comparable conditions. The most common include event detection algorithms such as the plateau algorithm [13], prediction algorithms such as Holt-Winters [6], and clustering algorithms [11].

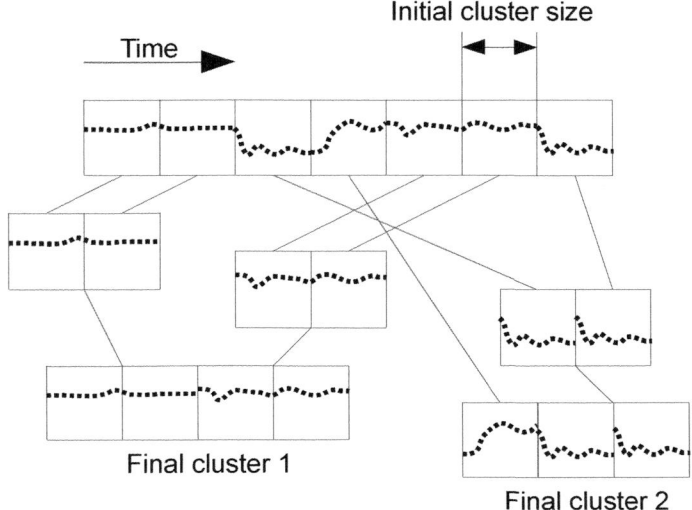

Fig. 1. Hierarchical clustering of 7 initial clusters into 2 clusters. The clusters with the smallest distance are combined in each step, regardless of the number of data points in the cluster. The clustering ends when the distance is too large.

We have deployed a *hierarchical, agglomerative* clustering algorithm to identify comparable periods. We start by dividing metadata into time series to form *initial clusters*, as illustrated in Figure 1. The number of consecutive data points used to form the initial clusters is called the *initial cluster size*. The initial clustering is a set of N small clusters. The algorithm starts by comparing the clusters, using a *distance metric* (measurement of how different two clusters are), to find the pair of clusters that are most similar among all possible pairs. The two chosen metadata clusters are then removed from the working set, merged into a single larger cluster, and this new cluster is added back to the working set, resulting in a working set of $N - 1$ clusters. The clustering process is then repeated until it reaches one of two termination conditions:

1. No more cluster pairs are found to be similar according to the distance metric. The limit for when clusters are considered similar is defined by a parameter *alpha*.
2. All data points have been combined into a single cluster.

Once the clustering of the metadata has terminated, the experiments are grouped accordingly based on the time stamps. This clustering algorithm is based primarily on the work by Wijk and Serlow [29], with the addition of the termination condition alpha. Pseudocode for the algorithm can be found in Figure 2.

```
while True do
  maxDistance := ∞
  for i in metaData do
    for j in metaData do
      if i ≠ j and maxDistance > distance(i, j) then
        maxDistance := distance(i, j)
        pair := (i, j)
      end if
    end for
  end for
  if maxDistance<alpha then
    (i, j) := pair
    metaData := metaData−{i}
    metaData := metaData−{j}
    combined := {i ∪ j}
    metaData := metaData + combined
  else
    return metaData
  end if
end while
```

Fig. 2. Clustering algorithm, with two parameters: *metaData* is a set of sets of metadata (divided according to the initial cluster size) and *alpha* is the cutoff value. Each data point in the metadata contains a timestamp and a measurement. Distance is a function that calculates the difference between two sets of metadata.

2.2   Distance Measurement and Clustering Parameters

We have multiple requirements for our distance measurement: it should be computationally efficient (as we work on relatively large data sets), it should handle the distributions commonly found in measurement data (long tails, binary distributions etc) and it must be able to compare inputs of different size.

We use the SciPy [9] implementation of the *Kolmogorov-Smirnov* (KS) 2-sample test [2] as a distance measurement. The algorithm performs well in the evaluations by Cottrell et al. of algorithms for network measurements [3]. One advantage with the KS 2-sample test is that it measures the similarity of sample sets without any assumptions about the distribution of the samples. It can compare sets of data of different size, and is efficient enough for our data sets.

For each clustering one (or several) metadata parameters are chosen to measure the similarity of network conditions. The choice of metadata parameters is highly dependent on the experiment, and requires insight into the properties of the particular experiments. In case the experiments are sensitive to multiple parameters, it is possible to either apply the algorithm separately on each parameter, or to use dimensionality reduction techniques [5]. An alternative approach could be to replace the KS test with a multivariate test such as the one described by Rosenbaum [21].

The two parameters for the clustering process—initial cluster size and alpha— may need tuning for the clustering to find as large sets of metadata as possible that represents similar conditions for the experiments in question. To achieve this, we use an approach where we vary the two clustering parameters to create multiple clusterings, and then use selection criteria to choose the most suitable clustering.

The parameter *initial cluster size* primarily controls resolution in the clustering process, and defines the shortest possible period of comparable conditions that can be identified. Variations that take place on a time frame that is shorter than the time corresponding to the initial cluster size are unlikely to be distinguishable. The KS 2-sample method performs well on small sample sets, compared to tests such as the χ^2 test [12]. Our tests indicate that the distance measurement does not perform well with less than 10 data points per cluster. In other studies, such as the one by Cottrell, up to 100 data points are used [3], and we use this as an upper bound on the initial cluster size. The upper and lower bounds for initial cluster size cannot be generally defined since they depend on several factors, such as length of the experiment and what type of other network activities that might disturb the experiment. These values probably need to be tuned differently for other experiments.

Alpha represents the limit for when two clusters of metadata are considered similar by the KS 2-sample test. The alpha value can be thought of as a way of controlling *sensitivity* in the distance measurement. The alpha value is normalized to the scale of 0.0–1.0. With a large alpha value, only large variations in the network conditions can be detected. With a lower alpha, the clustering process will be more discriminating, and hence result in more clusters. The level of variation in the network conditions is typically not known in advance. Therefore, our approach is to start with a high alpha and then run multiple clusterings for gradually decreasing alpha.

Running the algorithm several times with different parameters will result in a set of different clusterings. Each clustering will produce one or more clusters of metadata. The next step is then to select the most appropriate cluster. Since our goal is to identify the largest set of metadata that represents periods of comparable network conditions we need a selection criteria. We use SEM (Standard Error of the Mean) for this, and accordingly the cluster with the lowest SEM of the metadata is selected. By using SEM, the selection criteria takes into account both the number of samples and the variance of the samples within the cluster. The lowest SEM represents a large cluster with low variance in the metadata samples.

3   Evaluation in FEDERICA

As a way to evaluate our method we perform experiments in FEDERICA [26]. The experiments aim to study the effect of block size on transfer time for TFTP (Trivial File Transfer Protocol [23]). While the experiments are running we deploy background measurements to capture metadata, and introduce disturbances in the network.

The experiments consist of a basic protocol performance study: we measure the time it takes to transfer 100KB of data for different block sizes: 512 (the default for TFTP), 1024, 2048 and 4096 bytes. The purpose is to verify that our method can be used to obtain similar results from the disturbed and undisturbed experiments, which would serve as an indication that the method can be used to filter out periods of disturbances and thereby improve the quality of the results.

The experiment was deployed between FEDERICA nodes in Milan, Italy and Erlangen, Germany.

3.1   Metadata

We use latency measurements for metadata, since we expect our experiments to be sensitive to variations in latency. Hence, our hypothesis is that for this kind of measurements, latency is a suitable indicator for whether network conditions are similar. The background measurements are made with a slightly modified version of the Meter tool [8], originally developed for long term measurements on SUNET (Swedish University Network). Meter is designed for measuring latency and connectivity, using a set of tools including standard *ICMP ping* and a special *Latency* tool. Our metadata is obtained from the Latency tool, which estimates one way latency using a clock synchronization algorithm. Note that this tool is more accurate when it comes to changes in latency than absolute values (which is fine for our use of the metadata). We sanity check the data against latency measurements using ICMP Ping.

3.2   Disturbances

The FEDERICA nodes and links are dedicated for the experiment, so there are no competing activities that could influence our results. We introduce two types of load to disturb the network:

- Simulated high-definition video streams consisting of line-speed UDP traffic with packet size set to the MTU. This traffic is generated using pktgen [15], which is a software-based traffic generator. We denote this load *UDP*.
- TCP transfers of 100 MB per connection, called *TCP* load.

We schedule the disturbances by dividing the duration of the experiments into time intervals. For each interval, a randomized choice is made whether the interval should be disturbed. An interval is disturbed with probability $p = 0.25$ (and left undisturbed with probability $1 - p = 0.75$). Disturbances run continuously during disturbed intervals. Intervals are formed in two ways:

- All intervals are 30 minutes. We refer to this as Random scheduling.
- We generate events using a Poisson process with a mean arrival rate of one event per 30 minutes. The time between two successive events is an interval. We refer to this as Poisson scheduling.

The above gives four distinct combinations: UDP-Random, UDP-Poisson, TCP-Random and TCP-Poisson. We also run the experiments without disturbances (denoted Undisturbed data).

3.3   Executing the Experiment

We run the experiments for 6 hours for each type of disturbance, which results in 360 metadata samples and 1350 to 1430 data samples in each run. The order of the experiments (i.e., block size) is randomly chosen. For the clustering process, we vary initial cluster size between 10 and 100 data points and use 20 different alpha values in the interval 0.001 to 0.999.

3.4   Results

Table 1 shows the result of the clustering process. For our experiments, we would ideally get one cluster when we run the experiments without disturbances, and two clusters for configurations with disturbances—one cluster for the intervals without disturbances, and one cluster for the intervals with disturbances. Table 2 shows the results from the clustering process. The table reports the range of parameter values that give the best clustering according to the selection criteria (minimum SEM).

Table 1. Chosen clusterings by the selection criteria (minimum SEM)

	Alpha	Initial Size	Clusters	Size
Undisturbed data	0.001 – 1	10 – 100	1	357
UDP Random	0.001 – 0.4	25	2	257
UDP Poisson	0.001 – 0.4	10	3	184
TCP Random	0.001 – 0.2	20	2	254
TCP Poisson	0.001 – 0.2	15	2	239

For the undisturbed experiments ("Undisturbed data") the clustering process results in a single cluster no matter what parameter values are used. This is expected, since there are no disturbances and all experiments run at similar conditions. For the UDP disturbances, it is possible to distinguish disturbances up to alpha values of 0.4, while TCP disturbances are distinguished for alpha up to 0.2. Our explanation for the difference between UDP and TCP is that TCP disturbs the network less, due to the adaptive nature of TCP, and therefore TCP disturbances are not distinguished for higher alpha. For the initial cluster size parameter, we see that shorter initial cluster sizes are needed in order to distinguish Poisson disturbances. The reason for this is that Poisson disturbances can be more short-lived than Random disturbances, and therefore higher resolution is needed to identify them. Moreover, changing the initial cluster size changes the boundaries between initial clusters, and therefor we are unlikely to get exactly the same outcome for multiple initial cluster sizes.

Table 2 shows the results from the TFTP measurements. The results are shown for the experiments when no disturbances are introduced ("Undisturbed

data"), as well as for three different data sets from the experiments with disturbances. Data set 1 contains all measurement data, both disturbed and undisturbed. Data set 2 is obtained through the clustering process. Finally, data set
3 is the full data sets with the known disturbances removed. Ideally, both data
set 2 and 3 should give similar results to those from undisturbed data, and both
should yield better results compared to data set 1.

Table 2. Relative transfer time, normalized to 512 bytes block size. Mean values.

	C	Samples	BS=1024	BS=2048	BS=4096
Undisturbed data	1	357	50.43%	25.78%	13.55%
Data set 1: Full data sets					
UDP Random	1	347	52.49% (+2.06)	26.74% (+0.96)	15.29% (+1.74)
UDP Poisson	1	334	56.62% (+6.19)	28.43% (+2.65)	14.62% (+1.07)
TCP Random	1	356	51.08% (+0.65)	28.22% (+2.44)	14.66% (+1.11)
TCP Poisson	1	357	52.30% (+1.87)	25.78% (+0.00)	14.83% (+1.28)
Data set 2: Chosen cluster with comparable latency conditions					
UDP Random	2	257	51.69% (+1.26)	25.87% (+0.09)	13.57% (+0.02)
UDP Poisson	3	184	52.91% (+2.48)	26.80% (+1.02)	13.54% (-0.01)
TCP Random	2	254	49.42% (-1.01)	27.50% (+1.72)	14.05% (+0.5)
TCP Poisson	2	239	50.31% (-0.12)	23.97% (-1.81)	12.59% (-0.96)
Data set 3: Ideal clean data					
UDP Random	2	298	51.73% (+1.30)	26.26% (+0.48)	13.59% (+0.04)
UDP Poisson	2	237	52.22% (+1.79)	26.49% (+0.71)	13.56% (+0.01)
TCP Random	2	177	48.33% (-2.10)	25.54% (-0.24)	13.38% (-0.17)
TCP Poisson	2	190	50.27% (-0.16)	24.77% (-1.01)	13.01% (-0.54)

We observe that data set 1 differs more from the undisturbed data than set
2 and 3. So, both ideal clean data and our statistical method give more reliable
outcome of the experiment, compared to the unfiltered data. However, data set
3 differs less from the undisturbed data than data set 2 does, indicating that
there is still room for improvements.

This illustrates how we can identify comparable periods and thereby reduce
errors introduced by disturbances. In this case, our method improves the quality
of the experiment results, but does not identify all disturbances introduced.

4   Comparable Periods in PlanetLab

In the next step in the evaluation of our method we turn our attention to
PlanetLab, another popular experiment network. An accompaniment project,
CoMon [16], attempts to monitor a large number of host parameters in Planet-
Lab in order to aid individual researchers in identifying potential performance
problems that can impact experiments.

Unlike our experimentation in FEDERICA, we do not run experiments and collect measurement data in PlanetLab. Instead we investigate how our method can be used with CoMon data to characterize the performance of PlanetLab nodes. Furthermore, we investigate the performance of individual PlanetLab nodes, in contrast to the FEDERICA measurements that are end-to-end.

PlanetLab is a popular platform for experimental research, even though previous studies have found it hard to predict changes in PlanetLab [25]. Previous work by others have investigated the influence of competing activities on measurement results (see for example Whiteaker et al. [28], Wang et al. [27], and Sommers et al. [24]). Whiteaker specifically observes that using CoMon data to evaluate PlanetLab status (and in extension the experiment) can be valuable in improving experiment quality.

4.1 Parameters for the Clustering

CoMon regularly gather performance and configuration information from PlanetLab nodes, at a rate of around 12 samples per hour and node. The parameters monitored include debugging and performance information such as average load, CPU usage, memory usage, and network statistics. We examine CoMon data from January 1, 2011. We study two parameters: load average per minute and packet transmit rate (TX rate). The average load per minute, which is known to be a good measurement of aggregate CPU load on the system [4], has a resolution of two decimals. The TX rate is obviously important to any experiment that is sensitive to network load, and has a resolution of 1Kbps.

4.2 A Day in PlanetLab

The CoMon data we use is for 620 PlanetLab nodes. The resolution, the sampling rate, and the size of the metadata set limit the choice of clustering parameters. We vary initial clusters size from 60 minutes to 3 hours in steps of 10 minutes, which gives 12 to 36 samples per initial cluster. We examine eleven alpha values (0.001, 0.1, 0.2 – 0.9 and 0.999). Cluster selection is done by picking the cluster with minimum SEM, and for each of the 620 nodes, we investigate what cluster size gives lowest SEM. We take the result as the longest *comparable period* for that node, that is, the longest period over which conditions are comparable according to the metadata measure.

Using CPU load as metadata measure, we note that for 157 of the 620 nodes (just over 25%) the longest comparable period is the entire 24-hour period (as can be seen in Figure 3). For the remaining nodes, there are large variations, but with a tendency towards higher values. The mean of the longest comparable periods is around 16.56. We conclude that the nodes in PlanetLab have a tendency towards periods of comparable load conditions that are several hours long, and for many nodes the conditions do not vary much over time.

Examining the TX rate as metadata measure gives a different picture (Figure 3). Only around 5% of the nodes have comparable TX rate conditions over a 24-hour period, and for 27% of the nodes the longest comparable period found

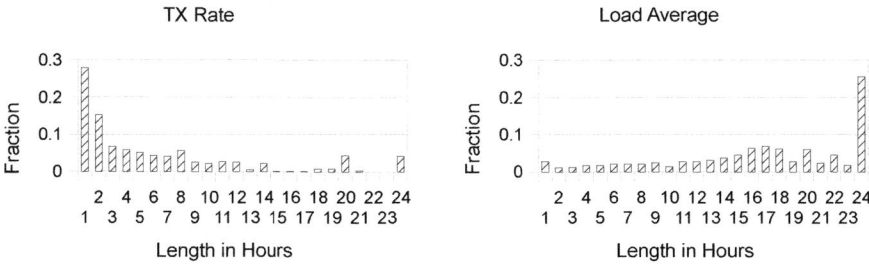

Fig. 3. Distribution of comparable periods in planetlab (2011-01-01), as chosen by the selection criteria among the clusterings. Note the difference in distribution between TX rate and Load Average.

is just one hour. This indicates that TX rate tends to change a lot more than CPU load over the day.

We conclude from the analysis that for a majority of the time, CPU load conditions on a given PlanetLab node are comparable, while TX rate exhibits comparable conditions for shorter periods of time. Whether these conditions are suitable for a specific experiment is a different question, since that would depend on the nature of the experiment. However, our experimentation indicates that researchers running experiments sensitive to changes in TX rate would have more reason for concern than those running experiments sensitive to variations in CPU load.

5   Conclusions

Clustering of metadata can help to identify periods of comparable conditions in network experiments. With this information it is possible to improve the quality of experiments in environments where conditions are not easy to predict. Our approach involves post-processing of metadata to group experiments into clusters such that experiments within the same cluster are performed at similar network conditions and are valid for comparison. Selection criteria are then applied in order to identify the cluster most suitable for analysis.

We investigate the use of Kolmogorov-Smirnov 2-sample method for clustering, and use SEM (Standard Error of Mean) as criterion for cluster selection. We evaluate the method by applying it on measurements in FEDERICA and PlanetLab. The results indicate that the approach work as intended, and can improve measurement results when experiments content for resources with other disturbing activities.

The clustering has two potential failure modes: It may fail to identify periods of comparable network behavior, in which case it is of no help to the user (but has no negative effect on the results either). The clustering may also fail by dividing experiments into multiple clusters even though there are no real differences in network conditions. This failure mode can be tested for to a certain degree, by examining whether there is a difference in experiment outcome between clusters.

The method has several evident limitations, which need to be taken into consideration: First, the method can only detect variations in network conditions that are captured by the metadata variables. Second, as any post-processing method that discards data, the method has to be used with care, since the discarded data could still have significance. Thirdly, the methods requires an understanding of the properties of an experiment and the sampling strategy, to avoid mistakes such as discarding a part of a sample. Finally, there is a risk that the process of collecting metadata interacts with the the experiment. This is a strong argument for passive collection of metadata.

Still, we believe that our proposed method is promising in several ways: It can be thought of as an informed way of cleaning data that also serves as a screening tool for data, gives better confidence in the environment where the experiment is executed, and helps researchers to identify variations in network conditions that might have impacted experiment results. Furthermore, unlike scheduling methods, there is no need to predict the conditions of the network.

Based on our two experiments we believe that our method can provide useful insights and improve experiment reliability. Still, several things remain: to study what background measurements give the best results for specific experiments, investigate different clustering algorithms, and design a tool for practical use of the method.

Acknowledgments. This work was financed by the FEDERICA project, a part of the EU Seventh Framework Programme for Research and Technological Development. We also wish to thank the anonymous reviewers for their valuable feedback.

References

1. Albrecht, J.: Achieving experiment repeatability on planetlab. In: NSF Workshop on Archiving Experiments to Raise Scientific Standards, NSF (2010)
2. Bickel, P.: A distribution free version of the smirnov two sample test in the p-variate case. The Annals of Mathematical Statistics 40(1), 1–23 (1969)
3. Cottrell, R.L.: Evaluation of techniques to detect significant network performance problems using End-to-End active network measurements. Technical report, NASA Center for AeroSpace Information (2006)
4. Ferrari, D., Zhou, S.: An empirical investigation of load indices for load balancing applications. Defense Technical Information Center (1987)
5. Fukunaga, K.: Introduction to statistical pattern recognition. Academic Pr. (1990)
6. Gelper, S., Fried, R., Croux, C.: Robust forecasting with exponential and Holt-Winters smoothing. Journal of Forecasting 29(3), 285–300 (2010)
7. Guralnik, V., Srivastava, J.: Event detection from time series data. In: Proceedings of the Fifth ACM SIGKDD International Conference on Knowledge Discovery and Data Mining, pp. 33–42. ACM (1999)
8. Hagsand, O.: Meter v2.2.2, http://www.nada.kth.se/~olofh/meter/ (accessed October 19, 2011)
9. Jones, E., Oliphant, T., Peterson, P.: SciPy: Open source scientific tools for Python (2001), http://www.scipy.org/ (accessed October 19, 2011)

10. Lee, S.-J., Sharma, P., Banerjee, S., Basu, S., Fonseca, R.: Measuring Bandwidth Between PlanetLab Nodes. In: Dovrolis, C. (ed.) PAM 2005. LNCS, vol. 3431, pp. 292–305. Springer, Heidelberg (2005)
11. Liao, T.W.: Clustering of time series data–a survey. Pattern Recognition 38(11), 1857–1874 (2005)
12. Massey, F.J.: The Kolmogorov-Smirnov test for goodness of fit. Journal of the American Statistical Association 46(253), 68–78 (1951)
13. McGregor, A.J., Braun, H.W.: Automated event detection for active measurement systems. In: Proceedings of PAM 2001 (2001)
14. Montgomery, D.: Design and analysis of experiments. John Wiley & Sons Inc. (2008)
15. Olsson, R.: Pktgen the linux packet generator. In: Linux Symposium 2005 (2005)
16. Park, K., Pai, V.S.: CoMon: a mostly-scalable monitoring system for PlanetLab. SIGOPS Operating Systems Review 40(1), 65–74 (2006)
17. Paxson, V.: Strategies for sound internet measurement. In: Proceedings of the 4th ACM SIGCOMM Conference on Internet Measurement, pp. 263–271. ACM (2004)
18. Peterson, L., Roscoe, T.: The design principles of PlanetLab. SIGOPS Operating Systems Review 40(1), 11–16 (2006)
19. Rahm, E., Do, H.H.: Data cleaning: Problems and current approaches. IEEE Bulletin on Data Engineering, 3 (2000)
20. Roscoe, T.: 33. The PlanetLab Platform. In: Steinmetz, R., Wehrle, K. (eds.) P2P Systems and Applications. LNCS, vol. 3485, pp. 567–581. Springer, Heidelberg (2005)
21. Rosenbaum, P.: An exact distribution-free test comparing two multivariate distributions based on adjacency. Journal of the Royal Statistical Society: Series B (Statistical Methodology) 67(4), 515–530 (2005)
22. Seibert, J., Zage, D., Fahmy, S., Nita-Rotaru, C.: Experimental comparison of peer-to-peer streaming overlays: An application perspective. In: Proceedings of the the 33rd IEEE Conference on Local Computer Networks, pp. 20–27 (2008)
23. Sollins, K.: RFC 1350: The TFTP protocol (Revision 2) (1992)
24. Sommers, J., Barford, P.: An active measurement system for shared environments. In: Proceedings of the 7th ACM SIGCOMM Conference on Internet Measurement, IMC 2007, pp. 303–314. ACM, New York (2007)
25. Spring, N., Peterson, L., Bavier, A., Pai, V.: Using PlanetLab for network research: myths, realities, and best practices. ACM SIGOPS Operating Systems Review 40(1), 24 (2006)
26. Szegedi, P., Riera, J., Garcia-Espin, J., Hidell, M., Sjodin, P., Soderman, P., Ruffini, M., O'Mahony, D., Bianco, A., Giraudo, L., et al.: Enabling future internet research: the federica case. IEEE Communications Magazine 49(7), 54–61 (2011)
27. Wang, G., Ng, T.: The impact of virtualization on network performance of amazon ec2 data center. In: 2010 Proceedings IEEE INFOCOM, pp. 1–9. IEEE (2010)
28. Whiteaker, J., Schneider, F., Teixeira, R.: Explaining packet delays under virtualization. SIGCOMM Comput. Commun. Rev. 41, 38–44
29. Wijk, J.J.V., Selow, E.R.V.: Cluster and calendar based visualization of time series data. In: Infovis, p. 4. IEEE Computer Society (1999)

tsdb: A Compressed Database for Time Series

Luca Deri[1,2], Simone Mainardi[1,3], and Francesco Fusco[4]

[1] Institute of Informatics and Telematics, CNR, Pisa, Italy
[2] ntop, Pisa, Italy
[3] Department of Information Engineering, University of Pisa, Pisa, Italy
[4] IBM Zürich Research Laboratory, Rüschlikon, Switzerland
{luca.deri,simone.mainardi}@iit.cnr.it
ffu@zurich.ibm.com

Abstract. Large-scale network monitoring systems require efficient storage and consolidation of measurement data. Relational databases and popular tools such as the Round-Robin Database show their limitations when handling a large number of time series. This is because data access time greatly increases with the cardinality of data and number of measurements. The result is that monitoring systems are forced to store very few metrics at low frequency in order to grant data access within acceptable time boundaries.

This paper describes a novel compressed time series database named tsdb whose goal is to allow large time series to be stored and consolidated in real-time with limited disk space usage. The validation has demonstrated the advantage of tsdb over traditional approaches, and has shown that tsdb is suitable for handling a large number of time series.

Keywords: Large-scale datasets, time series, web-based data visualization.

1 Introduction and Motivation

The demand of (near) real-time monitoring as well as the analysis of high-speed networks has put several constraints on monitoring systems. Users are demanding solutions able to interactively drill-down data while simultaneously collecting (hundred of) thousand metrics from various network sensors. In order to increase measurement accuracy, network administrators often reduce the sampling frequency of counters and gauges. If some years ago, a sampling period of 5 minutes was acceptable, nowadays network administrators require higher frequency samples for detecting anomalies that would not be detected by monitoring the same data at lower frequencies. For instance, detection of traffic spikes and microbursts require tenth (if not hundred) samples per second. The consequence of this trend is that monitoring systems produce an ever-increasing amount of data that needs to be stored and analyzed in a limited amount of time.

With the advent of multi-Gbit networks, the traffic being analyzed and the corresponding number of measured metrics increased significantly. Periodically accounting host traffic for a /24 subnet is very different from performing the same activity on a network backbone. In the latter case, monitoring systems cannot tolerate delays while

A. Pescapè, L. Salgarelli, and X. Dimitropoulos (Eds.): TMA 2012, LNCS 7189, pp. 143–156, 2012.

saving/reading data from/to the disk, as data access slow-down would prevent the system from carrying on the tasks within the expected timeframe.

As discussed in the following section, both relational databases and specialized tools such as the rrdtool [3], are used in the industry for handling *time series*. A time series is a sequence of data points measured at uniform time intervals (e.g. every 5 minutes) [16]. Unfortunately both solutions are only capable of satisfying requirements coming from small to medium environments, where the number of monitored metrics does not exceed a few tenth of thousand. However, collecting a much higher number of time series is not uncommon these days. Even a simple ntop installation [4] deployed for monitoring a medium network, has to keep track of several tens of thousand metrics just for keeping a few counters (e.g. bytes/packets sent/received). If additional counters are measured (e.g. traffic per protocol and network), the dataset size can quickly increase. According to the tests we carried on, existing open-source solutions and relational databases are affected by serious scalability issues when collecting hundreds of thousands (millions) metrics, making them *practically unusable* for large monitoring systems. In fact, without scalable and efficient time series handling tools, monitoring systems cannot store data at fine-grained granularity and are forced to decrease the monitoring accuracy. The lack of open-source tools for an efficient handling of time series has been the main motivation for creating a new open-source time series database for time series called *tsdb*.

The rest of the paper is organized as follows. Section 2 analyzes the various alternatives for handling time series. Section 3 presents the design and implementation of tsdb. Section 4 covers the tsdb validation and compares it with similar tools. Section 5 highlights some open issues and future work items.

2   Background and Related Work

2.1   Database Systems

For years network developers have used relational databases for storing network data persistently. Data with uniform characteristics are organized in tables linked by relationships. Each table is logically divided in columns and rows, where a row is uniquely identified by means of a primary key. Data stored on the database can be modified and deleted. Unfortunately network data is not characterized by many relationships, it is usually unchangeable (i.e. changing measured data might indicate a counterfeit), and the same data is repeated over time at every measurement interval, making relational databases not convenient for handling this type of data.

The reasons are manifold:

- At every measurement interval, tables are populated with fresh data that increases table cardinality. The consequence is that table cardinality as well the space taken on disk increases with the number of measurements.
- As soon as table indexes become large enough to prevent themselves to be cached, data retrieval becomes significantly slow [21, 14] thus jeopardizing the performance of applications sitting on top of the database.

A partial solution for avoiding these slowdowns is the use of binary large objects (BLOB) for storing time series. In this case the drawback is that the database is unable to search directly on blobs. As it can just read/store raw data, it must delegate to third party applications the implementation of such retrieval, and data management facilities. In order to address these issues with relational databases, time series database servers (TSDS) [1] have been created. They have been designed for enabling efficient data retrieval within some defined date/time ranges, as well for handling date and timezone conversions. Unfortunately TSDS are mostly used in the industry, and the only open-source alternative OpenTSDB [2], has a pretty complex architecture with several components interacting over a network, making it suitable only for distributed systems.

2.2   Round Robin Database

The Round-Robin Database (RRD) is a great alternative to relational databases for storing time series. It implements a file-based persistent circular buffer where data is stored according to its timestamp. When the database is created it is necessary to specify the data lifetime as well the frequency (named step in the rrd parlance) at which data is stored. For instance, it is possible to store a value every 5 minutes for at most 30 days. As all the information is specified at database creation, rrd files do not grow over time: their size is static and as large as the circular buffer. Each rrd database can store multiple time series, not necessarily all sharing the same lifetime and frequency parameters.

Typically rrd databases are small in size (64 KB or less) and stored as files on disk. Database files can be manipulated using a command-line tool named *rrdtool*, with no network access (e.g. via SQL-like query languages) typical of relational databases. Both rrdtool and its companion *librrdtool* library have been designed as tools to be accessed from the command line. Therefore, everything is file centric. Each database manipulation requires the library to open, manipulate, and save the file. If multiple operations have be performed on the same rrd file, the library needs to open/save/close the file multiple times. Another limitation is that most of the library functions require parameters to be specified in the argc/argv format, as required by the main() C function. The consequence is that due to these design shortcomings, it is not practically possible to define rrd databases with hundreds or thousands time series, thus limiting each rrd file to a few series.

This means that:

• The order of magnitude of rrd database files on disk is the same as the number of time series we need to handle.
• When we need to manipulate time series, we have to open/save/close as many files as the number of time series we plan to manipulate.

Although the rrdtool uses *mmap()* system call to reduce the number of read/write operations when updating an rrd or extracting a time series, the fact that rrd relies on the filesystem underneath is a matter of fact. Databases implement their internal data indexing and housekeeping, whereas in the rrd world this is delegated to the filesystem. Table 1 shows how rrdtool behaves while creating and updating a simple rrd archive (Ubuntu Linux 11.04 64 bit, Intel i7 860, SATA 3 Gb/s, ext4 filesystem). Note that in the RRD ecosystem, as the database is created as a circular buffer able to contain all defined time series, we use the term update (i.e. replace, if any, the previous value in the circular buffer) rather than the term append that is more appropriate for relational databases.

Table 1. rrdtool performance while creating and update rrd's

Number of RRDs	RRD Creation (Total)	RRD Update (Total)
10	0.53 sec	0.24 sec
100	5.34 sec	2.76 sec
1,000	58.13 sec	53.97 sec
10,000	600.74 sec	467.89 sec

The time spent per RRD is almost constant across all runs, with a limited increase with the number of files, probably due to disk management. We also performed additional tests using RAID, RAM disks, and various file systems (i.e. ext3, XFS, Oracle brtfs). Some setups reported better results with respect to Table 1, however the overall performance did not change significantly. This is because each rrd operation takes a few milliseconds; even if minimized, it needs to be multiplied for all rrds we plan to manipulate. This boils down to the conclusion that rrd is not able to manipulate a large number of time series within limited time boundaries. For example on our test system we have not been able to update more that 64k rrd's within 5 minutes time, although in our system no other application was accessing the disk and using CPU cycles.

The reasons for this behavior are manifold:

- RRD is file oriented, thus its performance cannot exceed the performance of the filesystem and disk it relies on.
- Each rrd file contains no more than a few time series, so at each time step all rrd databases need to be updated.

The rrd community has realized this limitation, and therefore on large installations a caching RRD daemon named *rrdcached* is used to cache updates in memory and perform them periodically. Even if this solution can reduce the number of rrd updates, it does not decrease the cost of per-rrd manipulation. Therefore, manipulating hundred of thousands (millions) rrds requires hours, making RRD unsuitable for effectively monitoring a large number of time series.

2.3   Additional Time Series Database and Tools

In addition to relational databases and rrdtool, there are other tools designed for handling time series such as TelegraphCQ [17], STATStream [18], and iSAX [19].

These tools had a large impact on research, but some of them are not maintained anymore, and others are just software prototypes used to validate the research work.

2.4   Numeric Databases and Compression

Compressing time series data can be fundamental not only for the obvious storage size reduction but also for improving performance. This is because less data needs to be read/written on disk. Suppose your data is stored in a block device (e.g. hard disk), reducing the data size by a factor of two, only half of the blocks will be read/written with respect to the same system when not using compression. The drawback is that compression has a cost in terms of time that is negligible if compared to the time required for moving mechanical parts (e.g. heads) of a hard disk.

Time series compression has applications in many network monitoring contexts where thousands if not million of monitoring metrics have to be collected at high frequency. The strategy to be chosen for compressing time series highly depends on the context. First of all, there are domains where a lossless compression is mandatory and others where an approximated representation of the time series is sufficient (lossy compression). The most widely used lossy time series compressors are based on concepts coming from signal theory and exploit specific properties, such as seasonality, to approximate the discrete signal represented by the time series of values. The rationale of these methods is to represent the signal in the frequency domain instead of the time domain to capture important signal properties such as periodicity. By capturing predominant patterns, lossy time series compression techniques enable operations such as nearest neighbor searches, and pattern searches [5] directly in the compressed domain, that are not possible with lossless compression techniques unless additional indexes are used. The main drawback of lossy compression techniques is that they rely on specific patterns for providing a good approximation of the given time series. This is the main reason why lossy compression has been rarely applied to network monitoring contexts, where the patterns of time series can drastically change due to anomalous events or to transient networking issues.

Lossless time series compression can be achieved by applying general-purpose lossless compression techniques, such as the popular Lempel-Ziv based compressors [6], or other compression techniques designed for better capturing and exploiting specific data patterns, such as the presence of long sequences of repeated symbols, or geometry distributions [7]. High-speed lossless compression of numerical values has been an active research topic during the last decade and has been driven by high-speed compression requirements coming from columnar databases [8] and information retrieval [9]. The main research focus has been on designing compressors optimized for achieving high compression and decompression speed rather than for achieving compression ratios as close as possible to the optimum. High-compression speeds, and more importantly decompression speeds are desired in all the contexts where the data has to be stored in a compressed form for reducing the volumes, but still frequently accessed for answering queries. If the compressors are able to provide a decompression speed that is higher than the read I/O bandwidth, then the compression is not only beneficial for reducing the volumes, but also, and more importantly for reducing the query response time. Among the family of speed-optimized compressors, Simple9 [9] and PFor [10] stand out for their performance. Both compressors achieve high decompression performance by optimizing the decompression routines for avoiding conditional branches, which are the bottleneck of current and future super-scalar processors. Compared to general-purpose compressors, the performance of high-speed integer compressors, both in terms of compression ratio and decompression speed, is more dependent on the data to be compressed. It has been shown that high-speed variant of LZ based compressors, such as LZO, have to be preferred over high-speed integer compressors, in case of data composed of many distinct numerical values [11].

3   tsdb: Design Goals and Implementation

As stated in the previous sections, both relational databases and tools like rrdtool are suitable for handling a limited number of time series as their performance is not satisfactory when the time series number increases (e.g. 100k or more). This has been the motivation for designing a new type of database named time series data base *(tsdb)* which is able to:

1. Handle millions of time series with minimal append and data extraction time.
2. Perform updates/appends on all the time series, as well as on a subset of them.
3. Add, remove and re-add time series as needed, without having to reconfigure or rearrange its structure.
4. Avoid data consolidation during append as in the RRD database. Thus it is possible to update/modify past data just by appending fresh data to the database, contrary to RRD where past data cannot be modified at all. If necessary, modification of existing data points can be prevented in tsdb.
5. Provide compressed data storage on a single file for all the time series. Backup and synchronization across network storage servers are easier without having to handle millions of files as with RRD database.
6. Support for time-limited series, so that the database will automatically purge data if older than the specified limit, similar to what rrdtool does.
7. Store time series in its native value without any aggregation (e.g. min/max/average) that might lead to loss of accuracy.
8. Support data extraction mechanisms for creating content suitable for web 2.0 applications that might plot data on a dynamic web page.
9. Provide simultaneous accesses to multiple applications, without writers blocking data readers.

Tsdb stores the database on disk and accesses it through a C API. It has been designed as a better RRD, able to handle millions of time series with limited disk space requirements and to feature fast data insert/retrieval for interactive use. For this reason both time series insertion and extraction have strict time requirements. Fast insertions are required to limit the latency between measurements and their consolidation into the database. High-speed retrievals are essential for minimizing the response time when, for example, web-based applications interactively read from the database Table 1 shows how the rrdtool speed limits the update time (e.g. it takes about 90 minutes to update 1 million series). This automatically implies the determination of a lower bound on the time interval between two consecutive measurements. The more time series are handled, the less often measurements can happen. For example, 90 minutes is the minimum time interval between two consecutive measurements in the previous case. Smaller time intervals would end up in determining buffer overflows.

As limiting data append time is critical, it is necessary to arrange time series so that their manipulation does not take too long. Databases and rrdtool arrange data on a per-time-series rather than on a per-measurement basis. This means that each individual rrd database contains all the values of a given series, so that at each measurement interval *all* the rrd database files have to be updated. Conversely, if data were arranged on a per-measurement basis, each update would have required the manipulation of a single rrd database. In a certain sense, rrdtool arranges data on disk in a way that is orthogonal to the more natural per-measurement way. It is worth noting that similar comments can be made for relational databases, where data is arranged per row (i.e. per time series) instead of per column (i.e. per measurement interval).

Changing the way data is arranged, significantly increases the database append speed but it has a negative impact on data retrieval. In fact whenever all the data points of a time series have to be retrieved, it is necessary to crawl across all the measurement intervals that fall within the time range of interest.

In tsdb all the time series stored in the same database share the same time intervals, so that all the series have the same number of data points and begin at the same time. For instance, all the series start October 1st 2011, and their samples are collected every 5 minutes. One one hand, this looks like a limitation with respect to other tools such as rrdtool where each rrd is independent. On the other hand, this is a nice feature for several reasons:

1. As all the series share the same time, comparing and extracting data points is simple and immune to time manipulation issues.
2. In real life, samples are collected at fixed-time intervals (e.g. once an hour) so it does not make sense to have different time points for series stored into the same database.

Having a uniform time across all the series, enables us to simplify the database design. tsdb has been developed incrementally. We started from a simple design, satisfying only a subset of the requirements discussed above and moved towards the current version, which satisfies them all.

In order to minimize the tsdb append time, we decided to arrange data per time interval. In early prototypes tsdb was implemented as a large extensible array, mapped on disk using the mmap() system call, with the time series in the rows and the measured values in the columns. The adoption of memory mapping minimized memory usage while creating a data structure large enough for storing all the data. Adding new measured values is straightforward as we simply need to add rows as in databases. This design guarantees very fast append time thanks to memory mapping and operating system caching. In fact data append basically happens in memory and a limited

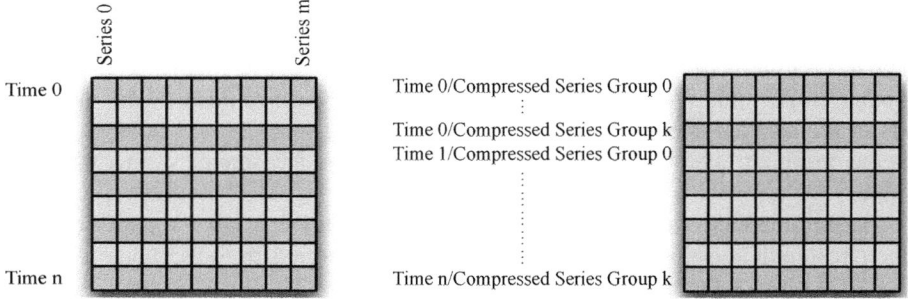

Fig. 1. (a) Time series arranged in an early tsdb prototype, (b) Compressed time series in the current tsdb implementation

slow-down is experienced only when the cache is flushed to disk. Concurrent operations are straightforward to implement, at least within the same process, as multiple threads see the database as a plain memory area. Inter-thread communication and synchronization can be avoided, unless the same memory locations need to be written by multiple threads. Unfortunately this design has also some drawbacks such as:

1. Data on disk is not compressed and thus its size increases linearly with the number of time series and data points.
2. Once the database is created, it is not possible to add/delete time series without rearranging the whole array.
3. Time series cannot be identified with a symbolic name but only by its index in the array, thus requiring an additional name to the index mapping facility.

In order to overcome these limitations, the current tsdb design is no longer a flat array but it is based on a key/value database used for:

1. Storing database metadata such as inter-measurement time interval duration (e.g. 1 hour), number of stored series and time of the last data update.
2. Associating time series name with a numerical index that is used to uniquely reference it inside the database. This mapping is more than a name-to-index association as it also contains the time ranges where such association is valid. For instance, suppose tsdb assigns index Y to a time series named X, added on January 1st 2011. If on March 1st 2011 X is dropped, the values until that day will be preserved into the database, but the index Y will be made available for new time series that might be later added to the database. If on June 23rd 2011, X is added again, a new index K will be assigned to it. Using the name-to-index mapping, tsdb is able to properly return all the data points values for X, according to the selected time interval, or the value 'undefined' for the intervals where X is undefined.
3. Storing time series according to the time interval. Suppose tsdb has to store values obtained from a measurement made on July 11th 2011 at 15:00. First, it converts the time into Unix epoch (1310389200 in the previous case). Then it creates a database entry identified by such epoch value. Tsdb is also responsible for normalizing the time according to the database configuration as well as for handling timezone conversion (i.e. all epochs are stored in GTM, Greenwich Mean Time).

The current tsdb database implementation is written in C and it is based on BerkeleyDB [13]. During the implementation of tsdb we have also explored other embedded databases such as GNU gdbm, but we have preferred BerkeleyDB as it is much faster when the database has several entries. Furthermore this database is resilient to crashes as it implements advanced techniques for preventing data loss, and it is commercially supported by a well-established database company such as Oracle. Contrary to many file-based databases such as SQLlite and gdbm, multiple clients can concurrently extract data. The tsdb library is implemented in about 1000 lines of code with data points represented as 32-bit integers. When data points for a new time interval are made available, tsdb creates an in memory array with an entry for each time series to be stored. For each time series it writes the corresponding measured value into the array. When the array is populated, tsdb stores it into the database using as key the epoch corresponding to the data. As motivated later in this section, data points for a given epoch are split in chunks of 10k elements and not saved all with the same epoch key. The key named convention we used is X-Y, where X is the epoch, and Y is the chunk id. For instance if we need to save 15k data points for epoch X, we will save the first 10k data points with key X-0, and the remaining 5k points with key X-1. In order to save space, prior to store data points, tsdb compresses and stores them in memory using QuickLZ [11] that we chose for its small memory footprint and high compression speed. As previously explained we did not evaluate the use of lossy compression as we are interested in storing exact values and completely avoid any approximation issues.

In addition to efficient storage of data points, it is essential to provide efficient data retrieval. As tsdb database records contain data points for a given epoch, when accessing data spanning across multiple time intervals, it would be necessary to decompress all the records falling within such interval before extracting the values. In order to minimize the amount of data to be decompressed and thus to reduce the data retrieval time, data points at a given epoch are grouped together and then saved into the database. From our tests, a good compromise between decompression speed and compression ratio is to group data points in chunks not larger than 10k elements. Even with a million of time series spanning one year and dumped at 1-hour intervals, the number of records is 876 k that is not a large number for the tsdb database.

In case most time series need to be analyzed, it is more efficient to convert a tsdb database in a format where all the data points of a time series are contiguous and uncompressed. In order to extract a single time series, we must decompress the chunk containing it for each time interval of interest (see Figure 1b). Assume each chunk contain data points for w time series (w=10k in the previous discussion). If the cost of decompressing a chunk is $O(w)$, we have that extracting all the w times series yields an amortized cost per series which is $O(1)$. We can apply this reasoning also to the whole tsdb database. If the total number of time series in the database is N, you have ceiling(N/w) chunks for each time interval. Since each chunk requires $O(w)$ time to be decompressed, the cost for decompressing a time interval of the whole database is $O(N)$ and consequently the amortized cost per time series is $O(1)$. However, the total cost for decompressing a whole database grows linearly with the number T of time intervals of interest since the previous operations have to be performed T times.

The tsdb database comes with a simple C API for creating applications based on it, as well as companion applications for storing and extracting data from the database, including a utility for extracting time series form tsdb and dumping them into an rrd database. This allows developers to use tsdb as a "faster rrd" while preserving backwards compatibility with rrd that is widely used in the network monitoring community. The following section explains how tsdb has been validated in real life, and compares its performance with similar tools.

4   tsdb Validation

tsdb has been created for effectively monitoring the Italian ".it" DNS registry, where two authors of this paper are currently working. In particular, we want to keep track of DNS queries for each individual ".it" Internet domain. At the time of writing such domains are more than 2.2 million. Before creating tsdb, we have used other databases such as MySQL and also advanced no_SQL [20] key-value databases such as Redis that natively supports the *set* data type, which we used for implementing time series. Both solutions perform reasonably well when handling a limited number of time series, but unfortunately their performance does not scale with the number of time series. For this reasons we have coded a few C test applications for comparing the various solutions. Figure 2 shows typical performance of the tools we have evaluated, when adding a variable number of data points on an empty database. It is worth mentioning that:

- With rrdtool and Redis, the number of database keys corresponds to the number of domains. Whenever new data points are added, the number of keys does not change as data points are added to the list of values associated with each key.
- Due to the nature of relational databases and thus of MySQL, whenever a new data point is added a new record with key <timestamp, key> is created, thus increasing the cardinality of table. This also happens with tsdb but at a much lower pace as records are packed in chunks of 10k records.

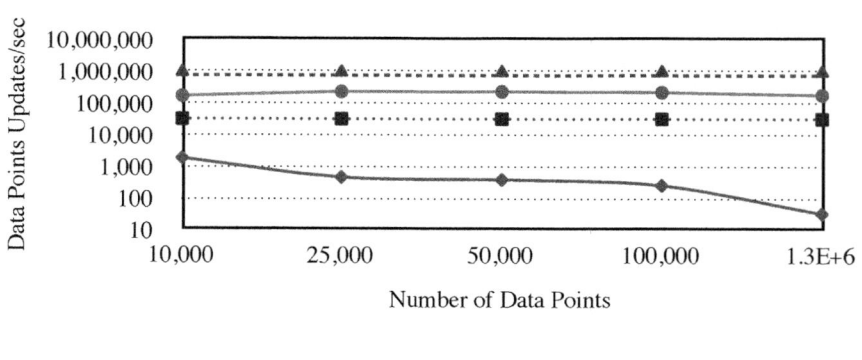

Fig. 2. Time series update processing speed (logarithmic scale)

The rrd performance started at 1600 updates/sec with small databases, and it decreases to 32 updates/sec after processing a few thousand records. We believe that the bottleneck is the disk subsystem, as rrdtool requires open/update/save for each rrd file. Redis performance is acceptable but still not comparable with the tsdb performance. In fact, this database is optimized for in-memory operations, so as long as there is free memory its performance is good. As soon as the available (8 GB) memory is exhausted (by Redis) the performance becomes poor (~ 30 updates/sec) and the system becomes due to disk swap. MySQL append performance of about 200k records/sec is only second to tsdb (that is almost 1M records/sec). Thanks to the use of QuickLZ, data compression ratio of data chunks is 1:5 in average with no noticeable performance degradation with respect to uncompressed data when appending data to tsdb. Storing fewer data has not just space but also performance advantages as previously discussed. For understanding disk space requirements of tsdb, we have created a monitoring application that saves in a tsdb database the number of queries for all ".it" DNS domains (approximately 2.2 million). Data retrieval is efficient, especially if compared with the performance offered by relational databases such as MySQL. To better position tsdb against MySQL, we have created two databases in both formats, containing the time series containing the time series for all .it Internet domains over a period of several months with daily measurements (i.e. one data point per domain per day). As we are monitoring the .it DNS since about 6 months, in order to simulate a larger dataset, we have replicated the data across several months. For tsdb we have used a single-file database for all tests instead of creating a database per year in order to create smarter queries that would only target a specific set of yearly databases. The MySQL table is defined as follows: `CREATE TABLE 'domains_summary' ( 'day' int(11) NOT NULL, 'domain_id' int(10) unsigned NOT NULL, 'num_queries' int(11) NOT NULL, UNIQUE KEY 'day' ('day','domain_id')) ENGINE=MyISAM DEFAULT CHARSET=latin1`. We have performed the same tests for both tsdb and MySQL with the exception for the largest dataset where only tsdb has been tested. In fact we believe that in this case the MySQL database would be partitioned based on the day, thus queries across several years would basically be as fast as the same query for a single year.

Table 2. tsdb vs. MySQL: (a) Database Size and (b) Search Time

Months of DNS Data	MySQL Database Size/Rows		tsdb Database Size/Keys	
6	8.2 GB	~ 300 M	0.8 GB	~ 41.6 K
18	25.0 GB	~ 900 M	2.6 GB	~ 125.6 K
69	-	-	6.3 GB	~ 497 K

Months of DNS Data	MySQL Search Time	tsdb Search Time
6	82 sec (40 sec)	2.21 sec (1.88 sec)
18	135 sec (124 sec)	2.47 sec (2.29 sec)
69	-	2.78 sec (2.31 sec)

In order to prevent data caching from affecting tests results, all search tests have been performed after a reboot. Table 2a compares the databases size. Thanks to data compression, tsdb is about 1:10 of the equivalent MySQL database. The use of chunks data aggregation allows the cardinality of tsdb keys to be greatly reduced. Please note that whenever Internet domains have no queries for a given day, in MySQL we do not insert a record (in tsdb the space is taken anyway although compressed at a higher ratio) explaining why the number of MySQL records is less than the number of tsdb keys multiplied by 10k (i.e. the chunk size). Table 2b shows the results when extracting 6 months of data for a single domain. Even on search time, tsdb performs dramatically better than MySQL, being also characterized by a lower pace when searching on larger files. In order to understand how cache affects results, between brackets we have put the time taken for immediately repeating the same query using a different domain name. We believe that in tsdb (a) the limited benefit of data caching, combined with (b) slightly flickering search time (i.e. searching two different Internet domains is not performed in the same amount of time) is due to data compression. This is because decompression times as well compression ratio are not completely uniform across data chunks.

Figure 3 shows a tool we created for exporting data from tsdb via JSON (JavaScript Object Notation) to a 2.0 web application that dynamically shows queries originated by a specific Autonomous System (AS) to the .it DNS servers. This has been possible only because the tsdb data extraction is so quick that we can perform it interactively on a dynamic web page.

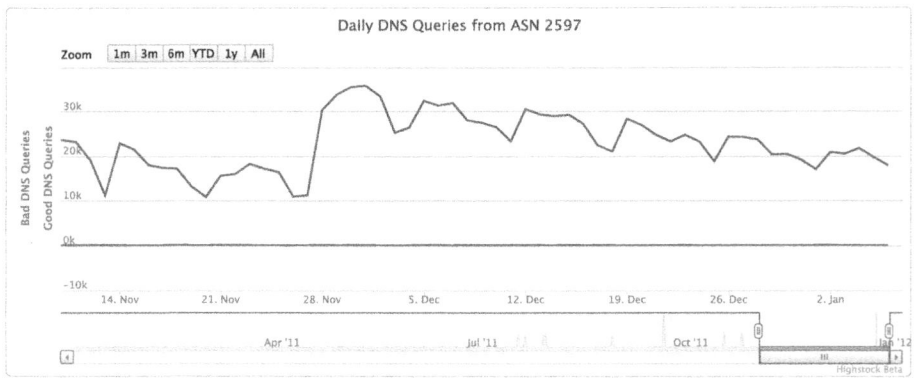

Fig. 3. tsdb time series export to a Web 2.0 interface

As previously discussed, tsdb does not have to be considered just as better tool for handling time series but rather as an enabling technology. One of the projects where we are working on at IIT/CNR focuses on monitoring the Internet AS-level structure. Within the scope of this project we have created a BGP (Border Gateway Protocol) client that connects to a border gateway and gets BGP updates from the router. Route changes are logged to file, and consolidated using custom tools we developed. So far we have used a relational database for tracking route changes. However we were not satisfied as its speed decreased over time due to the number of stored records. For this reason we migrated the system to tsdb where we maintain AS and route time series. This allowed us to both store data points at a much higher granularity (15 minutes in tsdb vs. one day in the relational database) and save disk space with respect to the previous solution. Having a higher granularity is a prerequisite for analyzing Internet topology changes and understanding its dynamics.

5   Open Issues and Future Work

So far tsdb is a library and a set of command line tools. We are planning to develop a tsdb server that can be remotely queried and instrumented through a network connection. We believe that the Redis query language is a good example of how to implement this in a simple and elegant way.

We are also trying to explore how tsdb could be enhanced for finding similarities between stored time series [15]. This is important for identifying Internet domains with overlapping behavior in DNS traffic, and autonomous systems with by similar route changes with the same time interval.

6   Final Remarks

Efficient time series handling is are very common requirement in many network monitoring contexts. Even if there is a significant amount of works described in literature, we have not been able to identify a tool for efficiently handling millions of time series in a simple yet effective way, and thus we have developed tsdb.

Its excellent performance when compared to similar tools and its simple design, makes it easy to integrate into existing applications. The validation phase has confirmed that tsdb can be effectively used for storing and analyzing a large number of time series in real life scenarios using limited disk space and characterized by append/search performances that are orders of magnitude better than the current state of the art. Finally, its backward compatibility with the popular rrdtool simplifies the migration of large existing infrastructures built upon rrdtool to tsdb.

Availability. This work is distributed under the GNU GPL license and it can be downloaded from https://svn.ntop.org/svn/ntop/trunk/tsdb/.

Acknowledgments. We thank Alfredo Cardigliano <cardigliano@ntop.org> and Alessandro Improta <alessandro.improta@iet.unipi.it> for their comments and valuable discussions during the design of tsdb.

References

1. Box, G., Jenkins, G.: Time Series Analysis: Forecasting and Control. Prentice Hall PTR (1994) ISBN: 0130607746
2. StumbleUpon, OpenTSDB: Open Time Series Database, `http://opentsdb.net`
3. Oetiker, T.: RRDtool: Round Robin Database Tool, `http://oss.oetiker.ch/rrdtool/`

4. Deri, L., et al.: Monitoring Networks Using ntop. In: Proc. of IM 2001 (May 2001)
5. Vlachos, M., Kozat, S., Yu, P.S.: Optimal Distance Bounds for Fast Search on Compressed Time-series Query Logs. TWEB 4(2) (2010)
6. Salomon, D.: Data Compression: The Complete Reference. Springer, Heidelberg (2000)
7. Rice, R.F., Plaunt, R.: Adaptive Variable-Length Coding for Efficient Compression of Spacecraft Television Data. IEEE Transactions on Communications 16(9), 889–897 (1971)
8. Harizopoulos, S., Liang, V., Abadi, D.J., Madden, S.: Performance Tradeoffs in Read-Optimized Databases. In: Proc. of VLDB (2006)
9. Anh, V.N., Moffat, A.: Inverted Index Compression Using Word-Aligned Binary Codes. Journal of Information Retrieval 8(1), 151–166 (2005)
10. Zukowski, M., et al.: Super-Scalar RAMCPU Cache Compression. In: Proc. of International Conference on Data Engineering, ICDE (2006)
11. Abadi, D., Madden, S., Ferreira, M.: Integrating Compression and Execution in Column-Oriented Database Systems. In: Proc. of 32nd ACM SIGMOD International Conference on Management of Data (2006)
12. Reinhold, L.M.: QuickLZ (2011), http://www.quicklz.com
13. Olson, M., Bostic, K., Seltzer, M.: Berkeley DB. In: Proc. of Usenix Annual Technical Conference (1999)
14. Mullins, C.S.: Database Administration: The Complete Guide to Practices and Procedures (2002) ISBN: 0-201-741296
15. Rafiei, D., Mendelzon, A.: Similarity-based queries for time series data. In: Proc. of the ACM SIGMOD (1997)
16. Brillinger, D.R.: Time Series: Data Analysis and Theory. Society for Industrial and Applied Mathematics (2001) ISBN-10: 0898715016
17. Krishnamurthy, S., et al.: TelegraphCQ: An Architectural Status Report. IEEE Data Engineering Bulletin 26(1) (March 2003)
18. Zhao, X.: High Performance Algorithms for Multiple Streaming Time Series, Ph.D. Dissertation, New York University (2006)
19. Shieh, J., Keogh, E.: iSAX: Indexing and Mining Terabyte Sized Time Series. In: Proc. of ACM SIGKDD (2008)
20. Tiwari: Professional NoSQL. John Wiley and Sons (2011) ISBN: 047094224X
21. Zaitsev, P.: Why MySQL could be slow with large tables? (June 2006), http://www.mysqlperformanceblog.com

Flexible High Performance Traffic Generation on Commodity Multi–core Platforms

Nicola Bonelli, Andrea Di Pietro, Stefano Giordano, and Gregorio Procissi

CNIT and Università di Pisa, Pisa, Italy

Abstract. Generating high-volume and accurate test traffic is crucial for assessing the performance of network devices in a reliable way and under different stress conditions. However, traffic generation still relies mostly on special purpose hardware. In fact, available software generators are able to reproduce rich and involved traffic patterns, but do not meet the performance requirements that are needed for effectively challenging the device under test. Nevertheless, hardware devices usually provide limited flexibility with respect to the traffic patterns that they can generate. The aim of this work is to design a traffic generator which can both achieve good performance and provide a flexible framework for supporting arbitrary traffic models. The key factor that enables our system to meet both requirements is parallelism, which is increasingly provided by modern commodity hardware: indeed our generator, which includes both kernel and user space components, can efficiently scale with multiple cores and multi–queue commodity network cards. By leveraging such a design, our generator is able to produce close-to-line-rate traffic on a 10Gbps link, while accommodating multiple traffic models and providing good accuracy.

1 Introduction and Motivation

As the global internet is increasingly becoming a critical infrastructure, being able to test network segments and network components in order to assess their robustness plays a crucial role in ensuring the trustworthiness of the network infrastructure. In such a perspective, reliable tools that can generate a realistic and verifiable traffic load are needed in order to discover weaknesses and faults in the network infrastructure before they become a cause of major damages, and to assess the capability of the equipment to keep working correctly even in the worst possible condition. On one hand, flexibility is needed in order to be able to mimic very specific traffic patterns, while on the other hand an ever increasing level of performance is required. Indeed, as the overall amount of the traffic carried by the internet is growing exponentially, the capacities of the links deployed on production networks is rapidly increasing (10 Gigabits are becoming more and more common). Therefore, a reliable testing tool needs to be able to generate an ever–growing amount of traffic. In spite of that, high performance traffic generation is a surprisingly understudied topic in the research community.

In many cases, the task of generating traffic to assess the performance of prototypes is delegated to hardware commercial devices. In order to reproduce

A. Pescapè, L. Salgarelli, and X. Dimitropoulos (Eds.): TMA 2012, LNCS 7189, pp. 157–170, 2012.

different traffic models, such generators usually either provide a well bounded set of possible models (as in the Spirent AX 4000) or reproduce real traffic traces captured over the network (like Napatech). Both approaches show severe limitations: in the former there is an evident lack of flexibility, as the models are implemented in hardware and therefore there is no way of introducing new features. The latter approach, however, despite being more flexible, still has several limitations: first, relevant traffic traces are treated by the law as personal data and are hardly available for the purpose of testing (besides, storing such traces is officially forbidden). In addition, a real traffic trace is usually not suitable when there is the need of reproducing particular traffic patterns (e.g. the worst case scenario) which are not common in real traffic.

As opposed to the limited flexibility of hardware based generators, a broad range of software based generators are available, which usually allow a very easy configuration of the generated traffic stream (often with graphical interfaces) and are able to support a wide range of protocols. However, those generators usually provide relatively low performance: most of them are able to fill only a fraction of a 1 Gb link. The currently available software tools, in addition, were mostly designed to run on single core commodity machines and are not able to make the most of the latest commodity hardware.

Indeed, the evolution of commodity hardware is pushing parallelism forward as the key factor that can allow software to attain hardware-class performance while still retaining its advantages. On one side, commodity CPUs are providing more and more cores (the next-generation Intel Xeon E 7500 CPUs will soon make 10 cores processors a commodity product), with a complex cache hierarchy which makes aware data placement crucial to good performance. On the other side, server NICs are adapting to these new trends by increasing themselves their level of parallelism. While traditional 1Gbps NICs (such as the very common Intel pro 1000 cards) exchanged data with the CPU through a single ring of shared memory buffers, modern 10Gbps cards (such as those based on the Intel 82599 controller) support multiple queues: multiple cores can therefore receive and transmit packets in parallel. In particular, incoming packets can be demultiplexed across CPUs based on a hash function (the so-called RSS technology) or on the MAC address (the VMD-q technology, designed for servers hosting multiple virtual machines). The Linux kernel has recently begun to support these new technologies (the 2.6.32 kernel is already multiqueue–aware).

In this paper we propose a novel architecture for packet generation, which effectively leverages NIC and CPU parallelism in order to combine the flexibility of the software based generators with hardware–class performance. In particular, our system is modular and allows a user to plug in new traffic models by writing a very limited amount of code; the massively parallel nature of the underlying system is completely hidden to the model developer and handled by the framework. The ability of effectively leveraging parallel processors guarantees that our architecture will keep scaling with the newer generations of CPUs, thus being able to attain higher and higher rates, the bottleneck of the single processor being eliminated. Multi-core aware design and implementation allow our generator to produce as much as 13 million packets per second on a commodity server, thus almost reaching wire-rate on a 10Gb link. In addition, differently from most of the solutions which need modified device drivers, it achieves such performance with vanilla drivers.

2 State–of–the–Art in Packet Generation

One of the most crucial performance show-stoppers in packet generation is the capacity of the socket which is used for sending packets down to the network interface. Therefore, we will briefly point outs the limitations of the state–the–art solutions.

Packet transmission sockets. Most of the current software based generators use the PF_PACKET socket, which comes with the standard Linux distribution and also supports a memory-mapping mode that allows to increase performance. However, such a socket has been designed in a single–core perspective and shows some limitations with respect to the modern architectures.

By accurately delving into its source code, we found several bottlenecks that we will describe in the following. The most evident weakness of PF_PACKET is that it does not allow to select a specific hardware queue for transmission when used on top of multi-queue NICs; this results in thread serialization when multiple threads send packets on the same device (no matter if they share the same socket or not) and it clashes with the base purpose of queue parallelism. In addition to this, PF_PACKET is based on a per-packet system call, like most of the classic I/O mechanisms. Such a dated design represents a remarkable overhead. Although the cost of the system–call can be amortized by means of batch–transmission (recent kernels introduce sendmmsg), such a system–call is hardly useful for a traffic generator, as there is no way to specify the inter–departure times for packets in the batch.

Another bottleneck of PF_PACKET is represented by the copy_from_user function, commonly used to transfer the packet payload to the kernel, and whose overhead is well-known to be higher than that of a normal memcpy performed to a memory-mapped region. However, this limitation can be worked around by using the recently introduced memory mapped version of the socket. Furthermore, packets transmitted by this socket are not directly conveyed to the NIC device driver but go through a series of mechanisms which brings several additional overheads. In particular, packets are also sent to registered sniffers, and therefore toward other open sockets. Since there is no way for a PF_PACKET socket to be used exclusively for transmission, this results in a severe performance penalty, especially in a multi–core scenario, where several sockets are in-use for parallel transmission (i.e. one socket per thread). Furthermore, when packet transmission is performed, the Linux traffic control (TC) also comes into play. This, in turn, involves an additional overhead even if no TC egress class is specified for the network device. This forces the packet transmission to be performed asynchronously by a different context, impacting adversely on the precision of the transmission time.

Recently, different solutions for improving the efficiency of Linux networking I/O have been proposed. Such solutions do provide good performance and are usually able of saturating a link with minimum sized packets. However, their scope is essentially different from that of our work, as they do not come with an integrated framework for synthetic traffic generation and are based on heavily patched drivers.

In [1], the authors present Packetshader, an extremely well performing software router, which is built around GPU acceleration of computation intensive and memory intensive functionalities (such as address lookup). Also, it relies on a heavily modified driver which introduces several optimizations, such as using a reduced version of the socket buffer structure and preallocating huge buffers to avoid per–packet memory allocations.

Netmap [2], a BSD based project, integrates in the same interface a number of modified drivers mapping the NIC transmit and receive buffers directly into user space. Deri recently released a heavily patched driver to be integrated into the well known PF_RING architecture [3], which allows to reach wire speed both in generation and in transmission, when simple test programs are used. However, as reported in [4], integration of such a system with a non multicore–aware traffic generator brings a significant performance decrease (the maximum packet rate being about 6 millions).

Software based traffic generators. Several open-source tools for traffic generation on commodity PCs have been proposed over the years, most of them designed for the Linux Operating System. KUTE [5] (an evolution of the former UDPgen) is an UDP traffic generator which is designed to achieve high performance over Gigabit-Ethernet. It is based on a Linux kernel module that operates directly on the network device driver bypassing the Linux kernel networking subsystem. However its performance is reported to be low and the project has not been supported for some years. RUDE [6] is able to instantiate simultaneous patterns of traffic, but it does not provide any explicit support for extensible interfaces and is not suitable to work at high rates, especially with small frames, as shown in [7].

MGEN provides both a command line and a GUI for user-friendly traffic generation in user-space. It runs on different Operating Systems such as FreeBSD, Linux, NetBSD, Solaris and Windows but its accuracy is limited by the system timers it is based on (e.g.: in the Linux kernel on PC-platforms, the timer resolution used by MGEN is only 10ms [6]). The Internet Traffic Generator (ITG) [8,9] aims at reproducing TCP and UDP traffic and replicate appropriate stochastic processes for inter–departure time and packet size. It is based on several distinct processes that are connected through Inter Process Communication and can actually support parallel generation. It is able to achieve a performance level comparable to that of RUDE and MGEN but provides more traffic patterns and runs also under Windows^{TM}. A version thereof [10] has been proposed for distributed measurements, but the authors reports its generated traffic to be below 650 Mb/s. Brute [7] is probably the best–performing software based traffic generator among the currently available ones and has an extensible modular architecture. However, as the other competitors, its design does not take parallelism into account and, therefore, it cannot scale properly on multi–core platforms. In particular, we measured its peak rate to be around one million packets per second. Ostinato [11] is a very recent traffic generator, which enables very flexible definition of the traffic flows via a graphical interface. However, its performance is reported to be quite poor when it comes to generating high traffic rates [12]. As already mentioned, [4] shows that, even when the packet transmission bottleneck is removed by using a properly modified driver, the design of Ostinato (which is not multi–core aware) limits the overall obtainable performance. Pktgen [13] is a software based traffic generator that runs within the Linux kernel directly, thus avoiding the overhead of communicating with a user space application. However, such a design choice limits its flexibility, as the range of traffic patterns it can generate is fixed and quite limited.

[14] gives an interesting insight about the level of performance and accuracy software based packet generators can reach. In particular, it compares MGEN, ITG and RUDE and shows that they cannot comply with the requested packet rate even at reasonably low speeds (well below 1Gbps). The paper also gives an interesting insight about how the OS scheduling interferes with packet generation and suggests polling as a way of improving both accuracy and performance (in fact, this is the solution we use in our architecture). [15] investigates further the correlation between OS scheduling and traffic generation accuracy.

Our approach. Our approach, whose cornerstone is the integrated co-design of kernel–space and user–space components, shows several significant contributions. In particular, with respect to the latest generation sockets:

- it adds a modular architecture which allows to transparently leverage parallelism for generating arbitrary traffic models. This is not trivial because, as reported, a non–aware design of the user space portion can cause a huge performance loss even with highly efficient sockets;
- it allows to attain high rates with non modified drivers, thus being able to cover a much broader range of hardware platforms.

With respect to user–space only traffic generators, instead, our design allows to attain much better performance, while still retaining a total flexibility with respect to the traffic models that can be generated. In particular, we propose an open architecture that users can extend to fit their needs.

3   The Generator Architecture

Our modular architecture is made up of several components. First of all, a set of parallel traffic transmitters is in charge of actually sending the packets to the NIC. Such transmitters are implemented as a novel socket named PF_DIRECT, along with an active context implemented as kernel–space thread. The context is in charge of either managing a specific device, or a hardware queue (for devices supporting them – notice that most of modern 10G NICs are equipped with

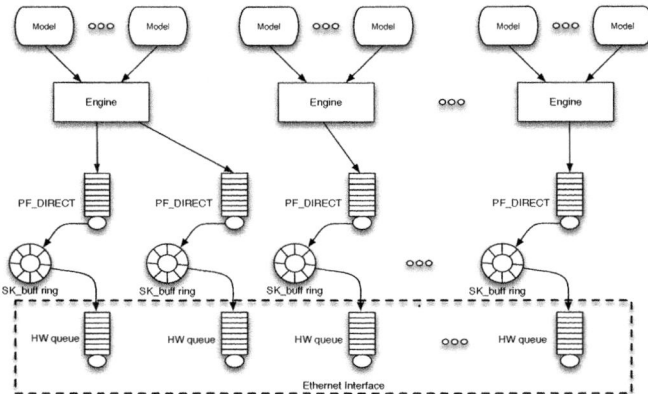

Fig. 1. Traffic generator architecture

hardware queues). PF_DIRECT sockets are fed with data to send by a set of traffic generation engines, which represent the user space threads that generate the global traffic streams out of a set of independent models. From a certain perspective, an engine is nothing but a discrete event simulator, which keeps ordered and updates the events generated by the traffic models; however, in order for the transmitters not to run out of data, the engine has real time requirements.

The traffic models are plug-ins that generate an ordered sequence of packets; they can set at will both the inter–departure times and the packet payloads, thus leaving the maximum freedom for implementing new models. In order for the system to avoid any data contention (which would impact on the performance), the described entities are associated into independent groups (user–space threads): each engine has a separate set of models to handle and a separate set of transmitters to feed. Notice that models served by the same engine are guaranteed to be strictly ordered with respect to the inter–departure times of packets (unless multiple hardware queues are used by the same engine, which may involve occasional reordering). Instead, models associated with different engines are independent and their synchronization relies on a common timer only. This architectural constraint is required to avoid the high cost of handling a shared resource among multiple cores. However, we point out that this is perfectly acceptable: as a use case, an engine could be serving a set of models which create an attack pattern, while a parallel one could be in charge of simulating a background traffic. Our architecture leaves a high degree of freedom in choosing different configurations, which can be specified as XML files.

3.1   Traffic Transmitters: PF_DIRECT

As already discussed, the Linux kernel places several bottlenecks on the path between the user space application and the NICs. In order to avoid this limitation we developed a novel socket, named PF_DIRECT, which allows efficient and scalable packet transmission.

The novel PF_DIRECT socket. The internal architecture of our socket is depicted in Figure 1 and is made up of the following components:

- a memory–mapped queue for payload and meta–data;
- a pool of pre–allocated socket buffers;
- a direct interface to a hardware queue.

We point out that a PF_DIRECT socket is bound to a network device or to a specific hardware queue and, being asynchronous, it has its own thread of execution in charge of transmitting packets. In addition to this, it is supposed to perform active waits in order to precisely reproduce the inter–departure times generated by the models. The SPSC (Single Producer Single Consumer) queue is the communication channel between the application and the socket. The queue consists of a memory–mapped area and two indices, specifying the last read position (which is written by the engine context and read by the socket context) and last written position (which, conversely, is read by the engine and written by the kernel context). Such a simple implementation gives a two-fold advantage: on one hand it avoids the performance cost of a system call, on the other it provides a wait–free mechanism for data sharing. The queue is used to convey both the packet payload (which is generated by the models) and some associated meta-data: packet length and, most importantly, transmission time.

Notice that the timing of the whole system is based on absolute timestamp-counters (TSC) available as a 64 bits register in modern CPU. TSC provides a number of advantages: on one hand, reading the value of the TSC register is much quicker (a few cpu cycles) than any other call/mechanism for reading the current time; besides, working with absolute time–points allows an easier and flexible dispatching of packets across multiple cores, which can follow a precise ordering of transmission without requiring inter-core communications. On the other, absolute time–points optionally leaves to the socket the possibility to recover from a late transmission by anticipating the following one, which would not be possible if the inter–arrival policy were specified. However, for multiple cores to work with absolute times, a common source of clock is required. Most server–class Xeon processors, for example, support the INVARIANT_TSC capability, which guarantees the timestamp counters on the different cores to be consistent. In case such feature is not available, an initial calibration procedure can be used to compute per–core offsets: in particular we can have all of the cores actively polling for a given atomic variable, then immediately read their own timestamp counter. Of course this may involve a small calibration error and we plan to devise a correction method based on multiple subsequent measurements as a future work.

The second component of the socket is a ring of pre–allocated socket buffers (sk_buff). While [1] reports allocation and initialization of the sk_buff structure to be one of the main bottlenecks in the Linux networking subsystem, using such a structure is mandatory in order to work with vanilla device drivers. Therefore,

PF_DIRECT allocates a pool of such structures at initialization time and cycles through them at run time: in particular, the payload and part of its metadata are copied from the shared queue into a socket buffer of the pool (at a negligible cost). The usage count of such a socket buffer is then forcibly incremented before the transmission, so that it is not deallocated by the device driver once transmitted.

The ring of socket buffers is required to amortize the latency of the clean-up routine: commonly it is used in drivers to free the socket buffers, while, in the case of PF_DIRECT, it just notifies that such a buffer is ready to be reused. The last component of the PF_DIRECT socket is the transmission routine, a direct interface to the device driver: it allows to skip the traffic control machinery that, as already pointed out, has a negative impact on the precision of the inter–departures time of packets.

3.2   Traffic Engines

The traffic engines are completely user–space threads of execution whose main function is generating an ordered stream of packets from a model set and dispatching them across a set of PF_DIRECT sockets (packet transmitters). For this reasons each engine keeps the models ordered in a heap according to the scheduled transmission time; they continuously extracts the first model from such a heap (i.e. the one with the closest transmission time) ans push its associated packet into the queue of one of the transmitters. The selection of most appropriate transmitter can be made according to a number of different criteria; we recall here that the packet transmitter perform active waiting until the intended transmission time for the packet is reached. The implemented criteria are:

- round robin: this policy provides good load balancing in terms of number of packets and is very simple;
- affinity: as the transmitters are synchronized and transmission is based on an absolute time–point, the packets generated from the models should be transmitted in order (i.e. in the order they are popped from the heap). However, if transmission times are close to each other and packets are assigned to different transmitters, minor timing errors can result in packet reordering. If a traffic model can by no means tolerate reordering, it can specify an affinity value, so that all of its associated packets are assigned to the same transmitter and strict ordering is enforced.

As above mentioned, in this preliminary prototype we only implemented these two policies: the implementation and evaluation of more involved schemes are left to future work.

3.3   Traffic Models

The traffic models are the components which are in charge of defining the traffic flows generated. Models are in fact plug-ins conforming to a simple interface and can be easily added by the user through a factory pattern implemented in C++.

The interface is defined as an abstract base class. Such a base class provides three essential protected methods to the derived classes (i.e. the actual models): one for accessing a buffer where the data to be transmitted is stored, one for setting the packet length, and another one for defining the inter–departure time of the packet. For the former, notice that defining a new packet involves no data allocation: the model just needs to overwrite the fields that change on a packet by packet basis. As for the latter, the choice of having a model work in terms of relative inter–departure times is intended to provide the maximum degree of abstraction to the model developer; conversion to an absolute time–point expressed in clock cycles is automatically performed by the framework. In order to define a new model, a developer has to implement two simple methods: one intended for configuration of the model's internals (e.g. a Constant Bit Rate – CBR – model needs to be provided a rate value) and called at start-up time, one called to schedule the transmission of a new packet when the previous one has been dispatched to a transmitter. We point out that our simple interface allows to reproduce even complex traffic model (we are currently working on a TCP emulator that should mimic two state machines on a network path). However, due to its performance penalty, closed loop generation is not supported.

As the packet–by–packet update method is on the fast data path, it is up to the developer to make it as quick and efficient as possible: a slow method can result in the transmitter queues to become empty and, in turn, in the generated traffic stream to have huge gaps and bursts (if the recovery algorithm is enabled). If a model needs an unavoidable degree of complexity, it is a good configuration choice to segregate it on a specific engine, so that it can have dedicated resources and cannot interfere with the rest of the traffic.

4   Experimental Results

In order to assess both the precision and the performance of our traffic generator, we carried out a number of tests by using different traffic analysis tools. Our generator always ran on a server–class machine, whose cost is below 2000$, which can be reasonably considered to be a commodity platform. Such a machine comes with a 6 cores Intel X5650 Xeon (2.66 Ghz clock, 12Mb cache), 12 GB of DDR3 RAM, and an Intel E10G42BT NIC, with the 82599 controller on board. In order to test our system with the maximum degree of parallelism, we kept Intel Hyperthreading enabled, thus carrying out the experiments with 12 virtual cores. The server runs Linux with the latest 3.0.1 kernel and the *ixgbe* 3.4.24 NICs driver.

In order to analyze the traffic produced by our generator, we used both commodity and special purpose hardware. A first set of tests is performed with the Spirent AX4000 hardware based protocol analyzer, which provides high resolution packet timestamping but, unfortunately, is not equipped with a 10Gb interface, thus limiting the maximum traffic rate to 1Gbps. We used it in order to obtain a very reliable characterization of the inter–arrival times of our generated traffic at relatively low rates (under 1 Mpkts/sec). As for 10Gb measurements,

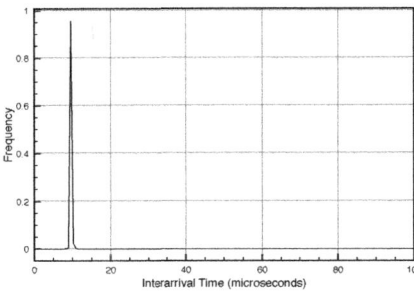
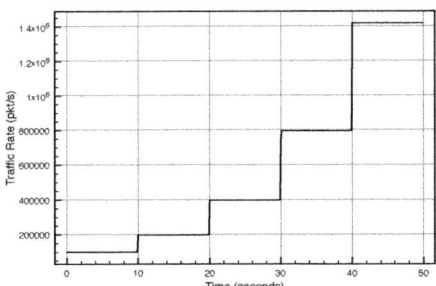

Fig. 2. CBR traffic – Rate: 100 kpkt/s   **Fig. 3.** CBR traffic – Increasing rates

we had to rely on a software solution running on a server which is identical to the one we used for generation. We measured the overall rate of the generated traffic by means of PFQ [16], a Linux kernel module designed for packet capture on multi-core architectures.

We captured the traffic generated by our system on a separate host (its hardware configuration being identical to the one of the generator); we also configured PFQ in order to timestamp the packet as soon as it was captured, in order to achieve the best resolution available on a software platform.

Overall, the performance evaluation here carried out will be further expanded in future works. In particular we plan to perform a more detailed investigation of the timing precision of our generator and to develop further its calibration mechanisms. In addition, the generator will be compared to a broader number of other generators on the same platform.

4.1   Up to 1 Gb/s Rates

The first set of tests aims at assessing the precision of the traffic generator in terms of packet inter–departure times and packet rates produced by CBR traffic models. We point out that in this section, unless it is explicitly remarked, we always used a single generation engine and a single hardware queue; also, generated packets are always 64 bytes long. Figure 2 shows the histogram of the packet inter–arrival times of CBR traffic at 100 Kpkt/s rate with a clear mode at 10 μs that, indeed, represents the constant inter–departure times of packet at that rate: in fact 99% of the samples lay in a range of 0.4 microseconds from the expected value. Figure 3, instead, is obtained by increasing every 10 seconds the generated packet rate, starting from 100 Kpkt/s up to line rate. The picture shows that the measured packet rate corresponds almost perfectly to the selected values for transmission; in addition, no significant rate fluctuation is reported, proving the high stability of the traffic generation process.

Figure 4 and 5 represent the histogram of inter–arrival times of two Poisson processes, with average rate of 100 Kpkt/s and 1 Mpkt/s respectively. The exponential behavior is clear in both cases (values of Figure 4 are plotted in linear scale while those of Figure 5 are reported in log scale) with values of rate

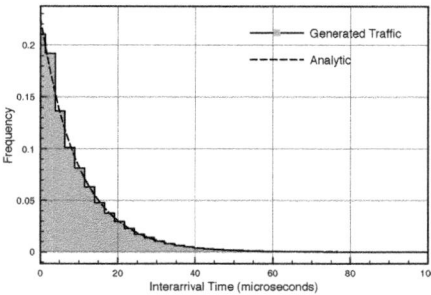

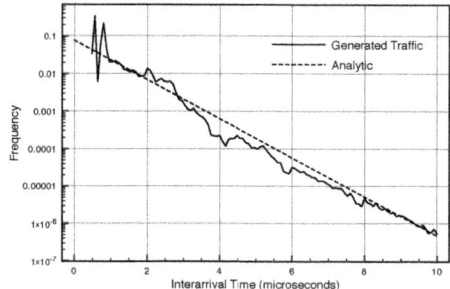

Fig. 4. Possion process – Rate: 100 Kpkt/s

Fig. 5. Poisson process – Rate: 1 Mpkt/s

consistent to those selected in generation. Only with small inter–departure times (under one microsecond) does the distribution in 5 show some spikes. This is most likely due to the quantized latency of the in–kernel polling cycle: indeed, this involves retrieving the timestamp value, which involves a non–negligible amount of clock cycles, thus leading to a discretized duration of the polling operation. The same phenomenon can be noticed in Figure 7.

Figure 6 reports the histogram of inter–arrival times of packets generated according to a Poisson process at 100 Kpkt/s rate by varying the number of hardware queues used in packet transmission. The histograms so obtained do not exhibit significant differences (they actually are almost perfectly overlapped), thus proving that the statistical properties of the traffic model are not affected by the number of active transmitters. This is an important result, as it shows that reordering of packets yield by the same model (which is theoretically possible when more than one hardware queue is used) is unlikely and does not affect the generator accuracy significantly.

Finally, to validate the overall architecture when multiplexing multiple traffic models, we compose together three independent Poisson processes with rates of 1 Kpkt/s, 4 Kpkt/s and 14 Kpkt/s respectively on the same generation engine. As well known from traffic theory, the resulting process must be again a Poisson process. Figure 7 clearly proves that the histogram of inter–arrival times is exponential (a straight line in log scale).

4.2   Towards 10 Gb/s Rates

In this section we report the results of several tests carried out at traffic generation rates higher than 1 Gbps. Figure 8 reports the histogram of inter–arrival times of a Poisson process with 4 Mpkt/s. rate (in this case we used multiple hardware queues to generate the distribution). As previously mentioned, in these cases measurements are taken by means of a software application (`pfq-isto`) running on top of PFQ: although we cannot make definitive statements on time precision, the figure shows a good exponential behavior. Notice that, unlike

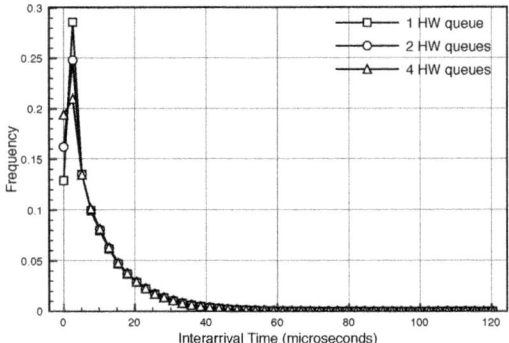

Fig. 6. Poisson processes generated with different number of HW queues (transmitters)

analogous tests carried out with lower rates, for the system to produce high packet rate multiple engines must be used on different cores. In the specific case shown in Figure 8 we used two engines each of them, in turn, feeding two transmitters (for an overall number of four transmitters involved in generating traffic).

As a result of the last set of experiments, Figures 9 and 10 show the cumulative amount of traffic that could be generated with our platform. To this end, we used a simple constant bit rate model and we adopted a configuration with 1 traffic engine and 4 packet transmitters (PF_DIRECT sockets); notice that in this case a single engine is enough to feed the transmitter due to the extreme simplicity of the CBR model, which does not involve random number computations.

Figure 9 reports the maximum cumulative packet rate that we can produce for different packet sizes. As we can see from the graph, our generator hits the line rate with 128 Bytes long packets and stays very close to it for minimum size (64 bytes) packets: indeed, it generates up to 13 million packets per second. The correspondent values in terms of bit rates are reported, instead, in Figure 10.

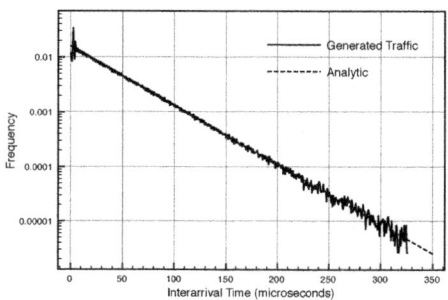
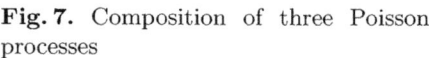

Fig. 7. Composition of three Poisson processes

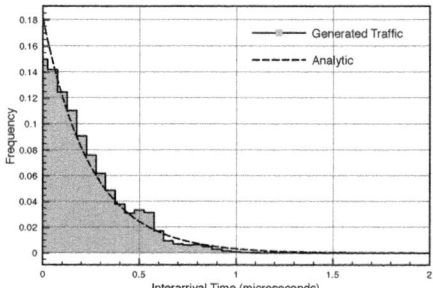

Fig. 8. Poisson process – Rate: 4 Mpkt/s

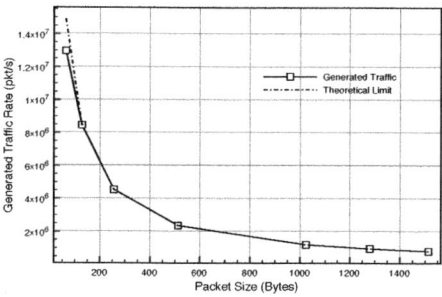

Fig. 9. Traffic packet rate vs. packet size

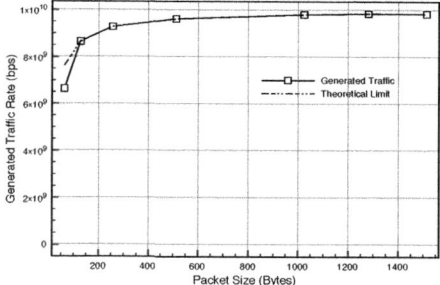

Fig. 10. Traffic bitrate vs. packet size

5   Conclusions and Future Work

In this paper we present a novel and flexible traffic generator specifically designed for multi–core architectures for the Linux operating system. The generator includes an accelerated transmission socket (named PF_DIRECT) that allows to reach multi–gigabit traffic figures by carefully taking advantage of computing parallelism and multi–queue hardware capabilities of modern 10 Gigabit NICs. In addition, the generator includes a set of pre–built traffic models but can be extended with any kind of user–defined additional model. From the performance point of view, the generator shows a good timing precision and hits the remarkable traffic generation rate of 13 Mpkt/s. with minimum sized packets while hitting line rate with packet sizes of at least 128 Bytes.

As a future work, we plan to perform a more thorough evaluation of the timing precision of our generator and to develop further its calibration mechanisms. We also plan to perform direct performance benchmarking by running other possible generators on the same platform. Furthermore, we envision to enrich its set of models by adding TCP emulation and others.

Acknowledgments. This work was partially supported by the Italian MIUR project IMPRESA and by DEMONS, a research project supported by the European Commission under its 7th Framework Program (contract-no. 257315). The views and conclusions contained herein are those of the authors and should not be interpreted as necessarily representing the official policies or endorsements, either expressed or implied, of the DEMONS project or the European Commission.

References

1. Han, S., Jang, K., Park, K., Moon, S.: Packetshader: a gpu-accelerated software router. In: Proceedings of the ACM SIGCOMM 2010 Conference on SIGCOMM, SIGCOMM 2010, pp. 195–206. ACM, New York (2010)
2. Rizzo, L.: http://info.iet.unipi.it/~luigi/netmap/

3. http://www.ntop.org/products/pf_ring/dna/ (2011)
4. http://www.ntop.org/pf_ring/building-a-10-gbit-traffic
   -generator-using-pf_ring-and-ostinato/ (2011)
5. http://caia.swin.edu.au/genius/tools/kute/
6. http://rude.sourceforge.net/
7. Bonelli, N., Giordano, S., Procissi, G., Secchi, R.: Brute: A high performance and extensibile traffic generator. In: Proc. of Int'l Symposium on Performance of Telecommunication Systems, SPECTS 2005 (July 2005)
8. Botta, A., Dainotti, A., Pescape, A.: Multi-protocol and multi-platform traffic generation and measurement. In: Proc. of INFOCOM 2007 DEMO Session (May 2007)
9. Avallone, S., Pescape, A., Ventre, G.: Analysis and experimentation of internet traffic generator. In: Proc. of New2an 2004, International Conference on Next Generation Teletraffic and Wired/Wireless Advanced Networking (February 2004)
10. http://www.grid.unina.it/software/itg/
11. http://code.google.com/p/ostinato/
12. http://code.google.com/p/ostinato/issues/detail?id=39 (2011)
13. http://lxr.linux.no/linux+v3.1.6/net/core/pktgen.c
14. Botta, A., Dainotti, A., Pescapé, A.: Do you trust your software-based traffic generator? Comm. Mag. 48, 158–165 (2010)
15. Paredes-Farrera, M., Fleury, M., Ghanbari, M.: Precision and accuracy of network traffic generators for packet-by-packet traffic analysis. In: Proc. of 2nd International Conference on Testbeds and Research Infrastructures for the Development of Networks and Communities, TRIDENTCOM 2006 (2006)
16. Bonelli, N., Di Pietro, A., Procissi, G.,
    http://netserv.iet.unipi.it/software/pfq/

Improving Network Measurement Efficiency through Multiadaptive Sampling

João Marco C. Silva and Solange Rito Lima

Departamento de Informática, Centro Algoritmi, Universidade do Minho, Portugal
pg15018@alunos.uminho.pt, solange@di.uminho.pt

Abstract. Sampling techniques play a key role in achieving efficient network measurements by reducing the amount of traffic processed while trying to maintain the accuracy of network statistical behavior estimation.

Despite the evolution of current techniques regarding the correctness of network parameters estimation, the overhead associated with the volume of data involved in the sampling process is still considerable. In this context, this paper proposes a new technique for multiadaptive traffic sampling based on linear prediction, which allows to reduce significantly the traffic under analysis, keeping the representativeness of samples in capturing network behavior.

A proof-of-concept, evaluating this technique for real traffic traces representing distinct traffic profiles, demonstrates the effectiveness of the proposal, outperforming classic techniques both in accuracy and data volumes processed.

1 Introduction

Traffic sampling techniques are extensively used to reduce the impact of performing traffic measurements on operational equipment. Their main objective is to select a subset of packets which will be used to estimate network parameters correctly, avoiding processing all network traffic. In adaptive sampling techniques the packet selection process may change dynamically during the measurement period, based on the value of a measured reference parameter. Despite the evolution on adaptive techniques correctness, their main target is not focused on reducing the overhead associated with the volume of data involved in the sampling process. This aspect directly impacts on monitoring costs and efficiency.

In this context, this paper presents a new multiadaptive traffic sampling technique based on linear prediction, which aims to reduce the volume of data handled in the network measurements without compromising estimation accuracy. For this purpose, the traffic selection process considers the levels of network activity, being configured to reduce measurement impact when the network activity increases and the measurement process tends to be heavier for measurement points. The multiadaptive behavior of this proposal is achieved considering both the sampling interval and the sample size as adaptive parameters, bounded by thresholds that guarantee the representativeness of samples in capturing network behavior.

The remaining of this document is organized as follows: related work is discussed in Section 2; the multiadaptive sampling technique is described in Section 3; the evaluation results are presented in Section 4; finally, the conclusions are drawn in Section 5.

A. Pescapè, L. Salgarelli, and X. Dimitropoulos (Eds.): TMA 2012, LNCS 7189, pp. 171–174, 2012.

2   Related Work

Common sampling adaptive techniques are usually based on Fuzzy Logic and Linear Prediction. In fuzzy logic techniques [2] [6], a controller adjusts the sampling rate based on past similar experiences, determining the most appropriate action for current traffic conditions [3]. This approach requires a long-term database to store the knowledge and the possible action for each situation.

Linear prediction based techniques [4] [5] try to forecast network behavior based on an observed parameter in past samples. In these techniques, when the prediction is correct, the sampling rate can be reduced, while inaccurate predictions indicate a change in network activity and, therefore, an increase in sampling rate is required to determine the new pattern behavior [3]. However, if the sampling frequency increases more resources will be required from the measurement point, precisely at the most critical moment of its operation.

The sampling technique proposed in this paper decreases resource consumption related to the processing, storage and transmission of captured packets in high network activity periods, maintaining the accuracy of network statistical behavior estimation.

3   Multiadaptive Sampling Technique

The multiadaptive sampling technique proposed in this work, grounded by previous research [3] [1], improves the sampling reactivity and efficiency, considering both the interval between samples and the sample size as adjustable parameters.

For this, the technique takes into account the last N samples to estimate the future value of the reference parameter. Thus, using linear prediction, the expected value Xp of the reference parameter for the next collected sample is defined by Equation 1 [3],

$$Xp = X[N] + \frac{\Delta T_{current}}{N - 1} \sum_{i=1}^{N-1} \left(\left| \frac{X[i+1] - X[i]}{\Delta T[i]} \right| \right) \tag{1}$$

where $X[N]$ is the most recente sample and ΔT is the time between the end of the sample $X[i]$ and the beginning of the sample $X[i+1]$.

When a new sample is collected, the corresponding value of the reference parameter S is compared with the expected value Xp. Based on a factor of change $m = \left| \frac{Xp - X[N]}{S - X[N]} \right|$ a set of rules is applied to define ΔT_{next} and ΔS_{next} used to schedule the start and the size of the next sample. The factor m returns a result close to 1 when the expected value Xp is close to the current value of S. The range of values that satisfies this condition is defined as $m_{min} < 1 < m_{max}$. If $m < 0.9$ the reference parameter is changing faster than predicted. This behavior indicates more network activity than expected. On the other hand, if $m > 1.1$ the value of the reference parameter is changing slower than predicted. However, if the value of S does not change since the last sample $X[N]$, m assumes an undefined value, which indicates that the network is stable. These values are set experimentally due to its good performance for multiple traffic types [3].

Table 1 lists the rules used to generate ΔT_{next} and ΔS_{next} from the current value of m. These rules reduce the amount of packets captured in high network activity periods.

Table 1. Rules to define the next interval between samples and sample size

current m	ΔT_{next}	ΔS_{next}
$m < m_{min}$	$\Delta T_{next} = m * \Delta T_{current}$	$\Delta S_{next} = \Delta S_{current} - (\Delta S_{current}/k), k = 4$
$m_{min} \leq m \leq m_{max}$	$\Delta T_{next} = \Delta T_{current}$	$\Delta S_{next} = \Delta S_{current}$
$m_{max} < m$	$\Delta T_{next} = \Delta T_{current} + 1seg$	$\Delta S_{next} = \Delta S_{current} + (\Delta S_{current}/k), k = 10$
$m_{undefined}$	$\Delta T_{next} = 2 * \Delta T_{current}$	$\Delta S_{next} = \Delta S_{current} + (\Delta S_{current}/k), k = 10$

An additional constraint limits the interval between samples and the sample size, guaranteeing a minimum number of samples to new predictions and avoiding to capture all traffic, respectively. These limits, tuned experimentally, define the minimum interval between samples as 0.1s and the maximum as 8s. The minimum sample size is defined as 0.1s and the maximum as 2s.

4   Evaluation Results

The proof-of-concept of the multiadaptive sampling strategy aims to determine its ability to capture the network behavior correctly with reduced overhead. Therefore, a statistical comparison of the reference parameter (network throughput with order of prediction $N = 3$) obtained through traffic sampling and the total traffic is carried out. In addition, the performance of the proposed strategy is compared to other commonly used techniques, the Systematic Time-based - ST technique (RFC 5475) and the Adaptive Linear Prediction - LP technique [3]. A set of statistical parameters is evaluated for four different traffic types, aiming at demonstrating the effectiveness of the proposal in distinct traffic scenarios. The *SIGCOMM 2008* scenario corresponds to a four hour capture of IEEE 802.11a communications (CRAWDAD). The *OC-48* scenario corresponds to a capture in a backbone link OC - Optical Carrier 48 (CAIDA). The *VoIP-SIP* represents around twenty hours of VoIP traffic, using SIP and G.711 encoding. The *Video Streaming* scenario corresponds to the capture of a live stream transmission in High Definition 720p encoded in MPEG-4 using RTP as transport protocol.

Table 2 shows that the multiadaptive technique (MA) leads to a significant overhead reduction for all traffic types under consideration, reducing up to 95% the data amount involved in the traffic characterization. The *data volume* corresponds to the sum, in Mbytes, of all packets collected with each sampling technique. The equivalence between the total traffic behavior and the behavior resulting from the sampling process is determined using *throughput* (kbps), *coefficient of variation* (CV), *correlation* and *relative mean error* (RME). CV measures the variability of time series, helping to characterize the burstiness of traffic. Correlation measures the statistical relationship between the sampled traffic and the total traffic. RME is used to assess the discrepancy between the mean of the total traffic and its sampled version [1].

From these results it is possible to verify that, despite the significant reduction on traffic volume considered, the multiadaptive technique ability to capture the real traffic behavior correctly is not compromised. Moreover, for several statistical parameters, the proposed technique outperforms the conventional techniques, which is even more significant when considering the high decrease on measurement overhead.

Table 2. Overall estimated behavior

Traffic/Parameter	Total	ST	LP	MA	Traffic/Parameter	Total	ST	LP	MA
SIGCOMM08					**SIP VoIP**				
Data volume (MB)	2382.92	395.97	511.93	120.49	Data volume (MB)	168.68	27.41	32.27	9.03
Throughput (kbps)	730.95	728.76	1130.27	803.65	Throughput (kbps)	21.85	21.30	50.49	22.03
RME		0.002	0.54	0.09	RME		0.02	1.31	0.008
CV	1.90	2.07	1.56	1.93	CV	5.28	5.41	2.90	5.08
Correlation		0.91	0.91	0.89	Correlation		0.71	0.88	0.86
OC-48					**Video streaming**				
Data volume (MB)	3189.67	533.39	454.92	165.72	Data volume (MB)	101.76	17.02	14.86	4.85
Throughput (kbps)	87103.35	87390.26	86467.86	86825.99	Throughput (kbps)	1275.84	1279.17	2998.53	1425.09
RME		0.003	0.007	0.003	RME		0.002	1.35	0.11
CV	0.10	0.09	0.10	0.13	CV	3.93	3.96	2.50	3.66
Correlation		0.84	0.81	0.82	Correlation		0.99	0.98	0.98

5   Conclusions

This paper has presented a multiadaptive traffic sampling technique based on linear prediction. The multiadaptive functionality is associated with the ability to change the interval between samples and the sample size according to the network observed activity. Controlling these two variables and defining appropriate bounds for their values, this technique reduces significantly the overhead associated with the traffic sampling process while maintaining the accuracy in estimating the statistical network behavior when taking throughput as reference parameter. Future work will be centred in further developing the technique regarding simultaneous estimation of parameters such as packet loss, delay and jitter, both in real time and in large-scale testbeds.

References

1. Dogman, A., Saatchi, R., Al-Khayatt, S.: An adaptive statistical sampling technique for computer network traffic. In: 2010 7th International Symposium on Communication Systems Networks and Digital Signal Processing (CSNDSP), pp. 479–483 (July 2010)
2. Giertl, J., Baca, J., Jakab, F., Andoga, R.: Adaptive sampling in measuring traffic parameters in a computer network using a fuzzy regulator and a neural network. Cybernetics and Systems Analysis 44, 348–356 (2008),
   http://dx.doi.org/10.1007/s10559-008-9005-0
3. Hernandez, E.A., Chidester, M.C., George, A.D.: Adaptive sampling for network management. Journal of Network and Systems Management 9, 409–434 (2001),
   http://dx.doi.org/10.1023/A:1012980307500
4. Lu, Y., He, C.: Resource allocation using adaptive linear prediction in wdm/tdm epons. AEU - International Journal of Electronics and Communications 64(2), 173–176 (2010)
5. Wei, Y., Wang, J., Wang, C.: A traffic prediction based bandwidth management algorithm of a future internet architecture. In: International Workshop on Intelligent Networks and Intelligent Systems, pp. 560–563 (2010)
6. Xin, Q., Hong, L., Fang, L.: A modified flc adaptive sampling method. In: WRI International Conference on Communications and Mobile Computing, CMC 2009, vol. 2, pp. 515–520 (January 2009)

Author Index

GPSR Compliance

The European Union's (EU) General Product Safety Regulation (GPSR) is a set of rules that requires consumer products to be safe and our obligations to ensure this.

If you have any concerns about our products, you can contact us on ProductSafety@springernature.com

In case Publisher is established outside the EU, the EU authorized representative is:

Springer Nature Customer Service Center GmbH
Europaplatz 3
69115 Heidelberg, Germany

Batch number: 09490872

Printed by Printforce, the Netherlands